SpringerBriefs in Mathematics

Series Editors
Nicola Bellomo
Michele Benzi
Palle Jorgensen
Tatsien Li
Roderick Melnik
Otmar Scherzer
Benjamin Steinberg
Lothar Reichel
Yuri Tschinkel
George Yin
Ping Zhang

SpringerBriefs in Mathematics showcases expositions in all areas of mathematics and applied mathematics. Manuscripts presenting new results or a single new result in a classical field, new field, or an emerging topic, applications, or bridges between new results and already published works, are encouraged. The series is intended for mathematicians and applied mathematicians.

More information about this series at http://www.springer.com/series/10030

Eduardo Garibaldi

Ergodic Optimization
in the Expanding Case

Concepts, Tools and Applications

Eduardo Garibaldi
University of Campinas - IMECC
Campinas, São Paulo, Brazil

ISSN 2191-8198   ISSN 2191-8201   (electronic)
SpringerBriefs in Mathematics
ISBN 978-3-319-66642-6   ISBN 978-3-319-66643-3   (eBook)
DOI 10.1007/978-3-319-66643-3

Library of Congress Control Number: 2017954979

Printed on acid-free paper

This Springer imprint is published by Springer Nature
The registered company is Springer International Publishing AG
The registered company address is: Gewerbestrasse 11, 6330 Cham, Switzerland

Preface

These notes grew out of a graduate course on ergodic optimization given by the author at the University of Campinas. Obviously, some background in ergodic theory is required to follow the text. The requisites are relatively light: a reader should be familiar with the basic concepts of ergodic theory, as contained in, for example, the first half of Walters' book [101]. Moreover, these notes are by no means meant to be exhaustive. As a matter of fact, we focus mostly on the interpretation of ergodic optimal problems as questions of variational dynamics (see, for instance, [40, 49, 51, 75]), in a comparable way to the Aubry-Mather theory for Lagrangian systems. The reader shall be conscious that other points of view are also useful in ergodic optimization, like the one based on properties of Sturmian measures and its generalizations (see, for example, [19, 26, 66]).

Ergodic optimization is a theoretical branch primarily concerned with the study of the so-called optimizing probability measures. The goal of this introductory monograph is, hence, twofold. One objective is to present and discuss in details fundamental concepts of the theory, in particular to clarify the relevance of a perspective dictated by the dual system:

(optimizing probabilities, sub-actions).

Sub-actions should be understood as the ergodic optimization analogue to subsolutions of the Hamilton-Jacobi equation. Therefore, another purpose of these notes is to provide pieces of evidence for the widest applicability of the viscosity solution methods or weak KAM solution techniques. There are several good books on this generalized notion of a solution of a partial differential equation, which describe its main properties and usual applications (see, for instance, [36, 45, 56, 78]).

There are a simple couple of reasons for choosing to discuss here ergodic optimal problems in an expanding context. First, the ergodic optimization theory on compact spaces has a more complete and detailed treatment for expanding dynamical systems. Hence, we decided to explain the essential theoretical aspects in a class of particular importance, namely, in symbolic systems, not only describing important achievements but also presenting some new results. Furthermore, we shall emphasize that the choice of a well-known dynamical framework allows making the exposition more intelligible, which is intended to be largely self-contained.

Campinas, Brazil   Eduardo Garibaldi
April 2017

Contents

Chapter 1
Introduction

In this opening chapter, we not only fix the dynamical system model on which these notes will be developed, but we also provide the core conceptual foundations of ergodic optimization theory. Moreover, we describe the structure of the entire text by anticipating the concepts and results that will be here covered, many of which have counterparts in Lagrangian Aubry-Mather theory. Our first step, nevertheless, consists in an attempt of placing ergodic optimization in the mathematical and physical research scenario.

1.1 Theoretical Interactions

Dedicated to gain a better understanding of optimal probabilities, the field of ergodic optimization has witnessed a major development over the last two decades, being nowadays an established branch of dynamical systems and ergodic theory. As a new area concerned with optimizing problems, it has received inputs from multiple domains, from dynamic programming to statistical physics, passing by transportation theory. As a matter of fact, part of the resulting research work is currently under way. Our intention is to point out at the level of a graduate textbook some of these productive theoretical exchanges, without pretense of listing all the relationships that have been explored so far by the academic community.

Lagrangian dynamics has been a major source of inspiration for ergodic optimization. In these notes, the contributions derived from such a research program play a prominent role. Since the seminal work of Contreras, Lopes and Thieullen [40], the usefulness and importance of examining the parallelism with Aubry-Mather theory became clear. This analysis has also served as a guide, highlighting key concepts that otherwise could have been taken by mere argumentative

© The Author(s) 2017
E. Garibaldi, *Ergodic Optimization in the Expanding Case*,
SpringerBriefs in Mathematics, DOI 10.1007/978-3-319-66643-3_1

device. We will present throughout the text a number of fundamental technical tools obtained from this perspective—for instance, the *Aubry set*, the *Peierls barrier*, the *Mañé potential*, and *separating sub-actions*. In this transfer of technology, influential articles on Lagrangian systems include, among others [12, 34, 37, 38, 44, 46, 80, 81]. At the end of this chapter, we provide precise information on the connections with Lagrangian Aubry-Mather theory that are covered in this book.

Zones of contact with other theories are also addressed in the text. Actually, the main theorem on the existence of separating sub-actions may be seen as a generalization of a solution for the problem of strict visualization scaling in max algebra (see [29]). To assist the reader to identify that tools in ergodic optimization can be interpreted as natural extensions of central notions in max algebraic (sub)eigenproblems, we shall discuss in these notes how the Mañé potential is related to the Kleene star. Introduced by a geometric series, the Kleene star allows to solve important class of linear equations in a complete idempotent semiring. One may find a short survey on methods and applications of max algebra in [54] and an exposition of how max algebraic questions and combinatorial optimization problems are strongly linked in [28]. For further information, see [8, 43, 53, 58, 59].

We will briefly shed some light on an interplay arising between ergodic optimization and solid-state physics. Significant studies [32, 33] on minimum-energy states and ground states of one-dimensional systems (including but not limited to the Frenkel-Kontorova model) are among the firsts to identify the strategic relevance of *calibrated sub-actions*. Derived from this physical context, the solution of the underlying fixed-point equation by means of iterates of the associated operator is the focus of Nussbaum's theorem [90] and is a nice example of how to find a sub-action for a locally constant potential. We decided to include this motivating example in these notes.

Ergodic optimization may be viewed as the zero temperature limit of thermodynamic formalism. This observation has guided a number of academic studies on the behavior of equilibrium states as the absolute temperature tends to zero. Considerable attention has been paid to the subject (see, for instance [5, 6, 10, 11, 14–16, 25, 30, 31, 42, 50, 61, 67, 69, 72, 84, 88, 100]). As this topic has been one of the most developed since the dissemination of the systematic analysis of optimizing probability measures, the closing chapter of this book provides a pedagogical introduction to this research theme from the influence of ergodic optimization theory. We will discuss a special case of Brémont's theorem [25] on the convergence of equilibrium states associated with locally constant potentials when the system is frozen.

We hope that the reader will benefit from the fact to be simultaneously exposed to parallel worlds. The bibliography should provide a sufficient start for those who may get interested in one of these areas that are in close contact with ergodic optimization.

1.2 Basic Notions

We shall now present in concréte mathematical terms the concepts that we need to introduce ergodic optimization theory in the expanding case. Given a continuous transformation $T : X \to X$ on a compact metric space (X, d), remember that the topological dynamics (X, T) is said to be expanding when there exists a constant $\lambda \in (0, 1)$ such that $d(x, y) \leq \lambda d(T(x), T(y))$ whenever $x, y \in X$ are sufficiently close. Most of the results we present in these notes could be immediately reformulated for any transitive expanding dynamical system. Since it is often easier to see many properties in symbolic systems first, we will consider here a symbolic dynamics setting. Nevertheless, the reader shall notice that some results will thus depend on the totally disconnect topology.

Main Notations *Let (Σ, σ) denote a one-sided topologically mixing subshift of finite type given by an $r \times r$ aperiodic transition matrix $\mathbf{M}$. In a precise way, $\mathbf{M}$ is a matrix with all elements either 0 or 1, for which there exists an integer $K_0 > 0$ such that all entries of the product matrix $\mathbf{M}^{K_0}$ are strictly positive. Besides, we are considering the subset of sequences $\Sigma = \{\mathbf{x} \in \{1, \ldots, r\}^{\mathbb{N}} : \mathbf{M}(x_j, x_{j+1}) = 1 \text{ for all } j \geq 0\}$ and σ is the left shift acting on Σ, $\sigma(x_0, x_1, \ldots) = (x_1, x_2, \ldots)$. For a fixed constant $\lambda \in (0, 1)$, we adopt on Σ the metric $d(\mathbf{x}, \mathbf{y}) = \lambda^k$, where $\mathbf{x}, \mathbf{y} \in \Sigma$, $\mathbf{x} = (x_0, x_1, \ldots)$, $\mathbf{y} = (y_0, y_1, \ldots)$ and $k = \min\{j : x_j \neq y_j\}$.*

We recall that, by Krylov-Bogoliubov theorem, the convex set $\mathcal{M}_\sigma$ of σ-invariant Borel probability measures is nonempty. Consider also $C(\Sigma)$ the space of continuous real-valued functions on Σ, equipped with the topology of uniform convergence induced by the maximum norm $\| \cdot \|_\infty$. So if we are given a continuous function $A \in C(\Sigma)$ called henceforth a *potential*, we can introduce the corresponding *ergodic maximizing value* $\beta_A := \sup_{\mu \in \mathcal{M}_\sigma} \int A \, d\mu$. Since $\mathcal{M}_\sigma$ is a weak* compact set and the map $\mu \in \mathcal{M}_\sigma \mapsto \int A \, d\mu \in \mathbb{R}$ is continuous, we actually have

$$\beta_A = \max_{\mu \in \mathcal{M}_\sigma} \int A \, d\mu. \tag{1.1}$$

In particular, it is easy to see that the mapping $A \in C(\Sigma) \mapsto \beta_A \in \mathbb{R}$ satisfies $|\beta_A - \beta_B| \leq \|A - B\|_\infty$ for all $A, B \in C(\Sigma)$.

Definition 1.A We denote by

$$m_A = \left\{ \mu \in \mathcal{M}_\sigma : \int A \, d\mu = \beta_A \right\}$$

the set of the A-maximizing probabilities.

The description of the above set is the main purpose of ergodic optimization. Note that m_A is a weak* compact set. Besides, by the ergodic decomposition theorem, m_A always contains at least one ergodic σ-invariant Borel probability. Actually, it is very easy to prescribe ergodic maximizing probabilities: if $\mu_i \in \mathcal{M}_\sigma$, $i = 1, \ldots, k$, are ergodic, notice that they are maximizing for the continuous potential $A(\mathbf{x}) = -\prod_{i=1}^{k} d(\mathbf{x}, \mathrm{supp}(\mu_i))$, where $\mathrm{supp}(\mu_i)$ denotes as usual the support of the measure μ_i. (For a more elaborated result, see, for instance [65].)

The other main notion we will discuss here may be also motivated by the ergodic maximizing value. In fact, thanks to duality, an important representation holds:

$$\beta_A = \inf_{f \in C(\Sigma)} \max_{\mathbf{x} \in \Sigma} [A(\mathbf{x}) + f \circ \sigma(\mathbf{x}) - f(\mathbf{x})]. \tag{1.2}$$

We will prove this equality in the next chapter. Meanwhile, we remark that a function that reaches the above infimum has a central role in ergodic optimization theory and therefore gives rise to the following key concept.

Definition 1.B A function $u \in C(\Sigma)$ is said to be a sub-action (for the potential A) when u verifies everywhere on Σ

$$A + u \circ \sigma - u \leq \beta_A. \tag{1.3}$$

For a first illustration of this concept, consider the potentials $-d(\cdot, \mathrm{supp}(\mu_i))$, with $\mu_i \in \mathcal{M}_\sigma$, $i = 1, \ldots, k$. Obviously their ergodic maximizing values are null. Note then that, for $u_n := -S_n\left(\prod_{i=1}^{k} d(\cdot, \mathrm{supp}(\mu_i))\right)$, where $S_n f = \sum_{j=0}^{n-1} f \circ \sigma^j$ denotes as usual the nth Birkhoff sum, we have $u_n \sigma - u_n = \prod_{i=1}^{k} d(\cdot, \mathrm{supp}(\mu_i)) - \left(\prod_{i=1}^{k} d(\cdot, \mathrm{supp}(\mu_i))\right) \circ \sigma^n \leq \prod_{i=1}^{k} d(\cdot, \mathrm{supp}(\mu_i))$. Since distances are bounded from above by 1, we get $-d(\cdot, \mathrm{supp}(\mu_i)) + u_n \sigma - u_n \leq 0$, $i = 1, \ldots, k$, which means that, for any $n \geq 0$, the continuous function u_n is a sub-action for all $-d(\cdot, \mathrm{supp}(\mu_i))$.

We will be interested in results for continuous sub-actions. However, notice that this regularity is not unreplaceable. After the inequality (1.3) which really characterizes a sub-action, its second more important property is the fact that the function $A + u \circ \sigma - u$ is integrable with respect to any σ-invariant Borel probability measure. So one could equally be interested in bounded measurable sub-actions. For more details, the reader may consult, for instance [83]. In Appendix A, one can have a glimpse of how to deal with bounded measurable sub-actions.

When a sub-action can be found, it gives important information on the maximizing measures. If a continuous function $u : \Sigma \to \mathbb{R}$ is a sub-action, we have a fundamental fact

$$m_A = \left\{\mu \in \mathcal{M}_\sigma : \mathrm{supp}(\mu) \subset (A + u \circ \sigma - u)^{-1}(\beta_A)\right\}. \tag{1.4}$$

Indeed, for the associated normalized potential $B := A + u \circ \sigma - u - \beta_A$, we note that $B \leq 0$ and $\int B \, d\mu = 0$ for all $\mu \in m_A$. Hence, when μ is an A-maximizing probability, $B \equiv 0$ μ-almost everywhere, which means that supp$(\mu) \subset B^{-1}(0) = (A + u \circ \sigma - u)^{-1}(\beta_A)$. Conversely, each σ-invariant Borel probability μ satisfying supp$(\mu) \subset (A + u \circ \sigma - u)^{-1}(\beta_A)$ must in a clear way be a maximizing measure.

Equality (1.4) explains why the sub-action concept is one of the essential tools in ergodic optimization. Therefore, it is not a surprise that the study of sub-actions has become a central matter. General properties of sub-actions in different dynamical settings can be found, for instance, in [19–21, 23, 24, 40, 41, 49, 51, 63, 64, 75–77, 86, 93, 96, 99].

We will dedicate several chapters to analyze specific properties of continuous sub-actions. Nevertheless, one should have in mind that, for a generic continuous potential, there does not exist even a bounded measurable sub-action. This fact may be seen as a corollary of a central result in [22] and a proof is provided in Appendix A. As one may immediately guess, the regularity of the potential plays a primordial role in the study of sub-actions. Here we will specially look for results relating to Hölder continuous potentials. Since we can simply incorporate the Hölder exponent into the distance, we remark that working with the Lipschitz class does not lead to loss of generality. Hence, if $f : \Sigma \to \mathbb{R}$ is Lipschitz continuous, as usual we set $\mathrm{Lip}(f) := \sup_{d(\mathbf{x},\mathbf{y})>0} |f(\mathbf{x}) - f(\mathbf{y})|/d(\mathbf{x}, \mathbf{y})$.

We will mostly focus on a distinguished subcollection of sub-actions.

Definition 1.C   A calibrated sub-action (for the potential A) is any sub-action $u \in C(\Sigma)$ such that, for all $\mathbf{x} \in \Sigma$,

$$u(\mathbf{x}) = \min_{\sigma(\mathbf{y})=\mathbf{x}} [u(\mathbf{y}) - A(\mathbf{y}) + \beta_A].$$

In these notes, calibrated sub-actions are mainly interpreted as discrete-time analogues of viscosity solutions of Hamilton-Jacobi equations. This perspective allows to carry out a systematic exposition that leads to a complete characterization of these sub-actions. Furthermore, as we will also see 'in the sequel, another interesting theoretical tool, namely, *separating sub-actions* may be derived from calibrated sub-actions.

1.3   Organization of the Text

The rest of this book is structured as follows. In Chap. 2, we present a proof for the dual characterization (1.2) of the ergodic maximizing value. By means of a Lax-Oleinik fixed-point iteration method, we guarantee the existence of Lipschitz continuous calibrated sub-actions for a Lipschitz continuous potential in Chap. 3. Besides, we show a result due to Nussbaum [90] which states that, for the particular case of a locally constant potential, a calibrated sub-action may be found after a finite number of iterations (see Theorem 3.4).

A subset of Σ plays a central role in the theory: the *Aubry set*. This set can be simply introduced as

$$\Omega(A) := \bigcap_{\substack{u \in C(\Sigma) \\ u \text{ sub-action}}} (A + u \circ \sigma - u)^{-1}(\beta_A). \tag{1.5}$$

Obviously, if we ensure the existence of sub-actions for the potential A, then from the set equality (1.4) we obtain that $\mu \in \mathcal{M}_\sigma$ is an A-maximizing probability if, and only if, its support lies on $\Omega(A)$. Another key concept is the *Peierls barrier*, namely, the function h_A defined on $\Sigma \times \Sigma$ by

$$h_A(\mathbf{x}, \mathbf{y}) := \lim_{\epsilon \to 0} \liminf_{n \to \infty} \inf_{\substack{d(\mathbf{z},\mathbf{x})<\epsilon \\ d(\sigma^n(\mathbf{z}),\mathbf{y})<\epsilon}} [-S_n(A - \beta_A)(\mathbf{z})].$$

We will show that $\{h_A(\mathbf{x}, \cdot)\}_{\mathbf{x} \in \Omega(A)} \subset C(\Sigma)$ is a family of calibrated sub-actions when A is Lipschitz continuous (see Proposition 5.2).

While Chap. 4 properly introduces the Aubry set, Chap. 5 is dedicated to detail the properties of the Peierls barrier as well as of other notion of action functional known as *Mañé potential*. We also show in Chap. 5 how the Mañé potential in ergodic optimization is strongly related to the Kleene star coming from max algebra. As already mentioned, not only is the terminology (Aubry set, Peierls barrier, Mañé potential, et cetera) borrowed from Lagrangian Aubry-Mather theory, but also some results get inspiration there. For instance, from [9, 49], we will see in Chap. 6 that any calibrated sub-action $u \in C(\Sigma)$ for a Lipschitz continuous potential A is characterized by its values on the Aubry set and the values of the Peierls barrier:

$$u(\mathbf{y}) = \min_{\mathbf{x} \in \Omega(A)} [u(\mathbf{x}) + h_A(\mathbf{x}, \mathbf{y})], \qquad \forall \, \mathbf{y} \in \Sigma.$$

This representation formula corresponds to the analogous one obtained by Contreras for weak KAM solutions of the Hamilton-Jacobi equation (see [34]). Other example of a result with similar predecessor in Aubry-Mather theory for Lagrangian systems is the existence of a *separating sub-action* $v \in C(\Sigma)$, namely, a sub-action satisfying $\Omega(A) = (A + v \circ \sigma - v)^{-1}(\beta_A)$. Separating sub-actions may be seen as a notion that is analogous to the critical sub-solutions of the Hamilton-Jacobi (see, for instance [46]). In Chap. 7, we discuss the existence of these sub-actions, more precisely, given a probability ρ on $\Omega(A)$ which is positive on all induced open sets, we prove that

$$v_\rho(\mathbf{y}) = \int_{\Omega(A)} h_A(\mathbf{x}, \mathbf{y}) \, d\rho(\mathbf{x}), \qquad \forall \, \mathbf{y} \in \Sigma,$$

is a Lipschitz continuous separating sub-action for A. The above expression may be interpreted as a generalization for a solution of the strict visualization scaling problem in max algebra (see [29]).

In Chap. 8, we show that, although the convex set of continuous sub-actions for a Lipschitz continuous potential in general is non-compact as a subset of the quotient space $C(\Sigma)/\mathbb{R}$, we can identify at least two extremal sub-actions. Moreover, we argue there that there are Lipschitz continuous separating sub-actions that do not admit the above convex combination representation.

Finally, we conclude these notes exhibiting some relevant relations between ergodic optimization and thermodynamic formalism. As already pointed out, we will be then interested in examples of the convergence of equilibrium states to a particular maximizing probability when the system is frozen (see Theorem 9.5).

Chapter 2
Duality

In this text, we use the dual characterization of the ergodic maximizing value to expose a natural bridge between maximizing probabilities and sub-actions. Since the earliest works in ergodic optimization theory, it became clear however that this constant can be presented in various equivalent ways. In this chapter, some alternative expressions that could be considered to define the ergodic maximizing value are brought to the attention of the reader. Furthermore, a proof of the dual formula is provided.

2.1 A Multifaceted Constant

There are several interesting descriptions of the ergodic maximizing value (1.1). For instance, as a topologically mixing subshift of finite type is an example of dynamics satisfying the specification property, from a classical result of Sigmund [97], the *periodic probabilities* (i.e., any σ-invariant Borel probability measure whose support is a periodic orbit) are a weak* dense subset of $\mathcal{M}_\sigma$. Actually, for symbolic systems, this fact is known at least since the work of Parthasarathy [91]. Therefore, one has

$$\beta_A = \sup \left\{ \int A \, d\mu : \mu \in \mathcal{M}_\sigma, \ \mu \text{ periodic probability} \right\}. \tag{2.1}$$

Naturally related to this formula, the rate at which the ergodic maximizing value is approached by integrals of the potential against periodic probabilities of at most a given period was fully studied in [27].

Birkhoff's ergodic theorem immediately suggests that the ergodic maximizing value (1.1) may be also expressed in terms of averages of Birkhoff sums, placing a major emphasis on orbits. Indeed, well-known characterizations of this constant are the following ones.

© The Author(s) 2017
E. Garibaldi, *Ergodic Optimization in the Expanding Case*,
SpringerBriefs in Mathematics, DOI 10.1007/978-3-319-66643-3_2

Proposition 2.1 *For a continuous potential A,*

$$\beta_A = \lim_{n\to\infty} \frac{1}{n} \max_{\mathbf{x}\in\Sigma} S_n A(\mathbf{x})$$

$$= \max_{\mathbf{x}\in\Sigma} \limsup_{n\to\infty} \frac{1}{n} S_n A(\mathbf{x})$$

$$= \max_{\mathbf{x}\in Reg(A)} \lim_{n\to\infty} \frac{1}{n} S_n A(\mathbf{x}),$$

where Reg(A) denotes the set of the points $\mathbf{x} \in \Sigma$ *for which there exists the limit of* $\frac{1}{n} S_n A(\mathbf{x})$ *as n tends to infinite.*

A proof of such equalities may be found, for instance, in [64]. For the sake of completeness, we present a similar argument.

Proof First note that $\{\max S_n A\}_{n\geq 1}$ is subadditive and thus the limit of $\frac{1}{n} \max S_n A$ exists (and is equal to $\inf_{n\geq 1} \frac{1}{n} \max S_n A$). Moreover, it is easy to see that

$$\sup_{\mathbf{x}\in Reg(A)} \lim_{n\to\infty} \frac{1}{n} S_n A(\mathbf{x}) \leq \sup_{\mathbf{x}\in\Sigma} \limsup_{n\to\infty} \frac{1}{n} S_n A(\mathbf{x}) \leq \lim_{n\to\infty} \frac{1}{n} \max_{\mathbf{x}\in\Sigma} S_n A(\mathbf{x}).$$

On the one hand, recalling that Borel probabilities form a weak* compact set, it is not hard to check that, for any given sequence $\{\mathbf{x}_n\} \subset \Sigma$, the measures defined by $\mu_n := \frac{1}{n} \sum_{k=0}^{n-1} \delta_{\sigma^k(\mathbf{x}_n)}$ always accumulate on σ-invariant probabilities. Hence, choosing $\mathbf{x}_n$ such that $S_n A(x_n) = \max S_n A$, we conclude that $\lim \frac{1}{n} \max S_n A \leq \beta_A$.

On the other hand, Peres' lemma (see [92]) states that, for any $\mu \in \mathcal{M}_\sigma$, there exists $\mathbf{x} \in \Sigma$ such that

$$\int A \, d\mu \leq \frac{1}{n} S_n A(\mathbf{x}) \qquad \forall n \geq 1.$$

Considering then a maximizing probability $\mu \in m_A$, we have

$$\beta_A \leq \liminf_{n\to\infty} \frac{1}{n} S_n A(\mathbf{x}) \leq \limsup_{n\to\infty} \frac{1}{n} S_n A(\mathbf{x}) \leq \lim_{n\to\infty} \frac{1}{n} \max S_n A \leq \beta_A,$$

which shows that $\mathbf{x} \in Reg(A)$ and therefore all equalities hold. $\qquad\square$

The dual formula (1.2), which will be addressed in a moment, corresponds in Lagrangian Aubry-Mather theory to a similar description of Mañé's critical value (see [38] for details). We provide in the next result a representation of the ergodic maximizing value using Mañé's viewpoint (compare (2.2)–(5.9) in Chap. 5). As usual, we denote by orb($\mathbf{x}$) the orbit by σ of a point $\mathbf{x}$ of Σ.

Proposition 2.2 *Let $A : \Sigma \to \mathbb{R}$ be a continuous potential. Then*

$$\beta_A = \min \left\{ c \in \mathbb{R} : \sum_{y\in orb(\mathbf{x})} (A - c)(y) \leq 0, \quad \forall \mathbf{x} \in \Sigma \text{ periodic point} \right\}. \qquad (2.2)$$

Proof Obviously $\int (A - \beta_A)\, d\mu \leq 0$ for all μ σ-invariant probability measure and, in particular, for all μ periodic probability. Moreover, if $\int (A - c)\, d\mu \leq 0$ for any μ periodic probability, thanks to (2.1), we deduce that $\beta_A \leq c$, which concludes the proof. $\qquad\qquad\qquad\square$

2.2 The Dual Formula

In ergodic optimization, the dual expression was established in two unpublished papers [41, 95] by different techniques. For convenience of the reader, we present a complete proof, which may be extended to other situations, like the one involving the study of minimizing configurations in almost-periodic environments [52]. A comparable argument was used in [13] to obtain the dual description of Mañé's critical value.

Theorem 2.3 *For a continuous potential* $A : \Sigma \to \mathbb{R}$,

$$\beta_A = \inf_{f \in C(\Sigma)} \max_{\mathbf{x} \in \Sigma} [A(\mathbf{x}) + f \circ \sigma(\mathbf{x}) - f(\mathbf{x})].$$

Proof Given any $f \in C(\Sigma)$, notice that, for all $\mu \in \mathcal{M}_\sigma$,

$$\int A\, d\mu = \int (A + f \circ \sigma - f)\, d\mu \leq \max(A + f \circ \sigma - f),$$

so we get $\beta_A \leq \inf_f \max_{\mathbf{x}} (A + f \circ \sigma - f)(\mathbf{x})$.

In order to obtain the opposite inequality, by conciseness we denote

$$\kappa := -\inf_f \max_{\mathbf{x}} (A + f \circ \sigma - f)(\mathbf{x}).$$

Notice that it is enough to prove the following claim.

Claim There is a measure $\mu \in \mathcal{M}_\sigma$ such that $\int g\, d\mu - \max(A + g) \leq \kappa$, for all $g \in C(\Sigma)$.

In fact, if such a claim holds, for $g = -A$, we have $-\kappa \leq \int A\, d\mu \leq \beta_A$, which is the inequality we are looking for. To demonstrate the claim, recall that a function $g \in C(\Sigma)$ is a (topological) coboundary when $g = f \circ \sigma - f$ for some $f \in C(\Sigma)$. Consider then the sets

$$C_1 := \{(g_1, t_1) \in C(\Sigma) \times \mathbb{R} : t_1 < 0,\ g_1 \text{ is a coboundary}\},$$

$$C_2 := \{(g_2, t_2) \in C(\Sigma) \times \mathbb{R} : \kappa + \max(A + g_2) < t_2\}.$$

Notice that C_1 and C_2 are nonempty convex sets. Besides, they are disjoint by the definition of κ and C_2 is open. Therefore, by the geometric version of Hahn-Banach theorem, there exists a non-null continuous linear form (ν, α) on $C(\Sigma) \times \mathbb{R}$ which separates C_1 and C_2:

$$\sup_{(g_1,t_1)\in C_1} \left(\int g_1 dv + \alpha t_1 \right) \leq \inf_{(g_2,t_2)\in C_2} \left(\int g_2 dv + \alpha t_2 \right). \tag{2.3}$$

We must have $\alpha > 0$. Indeed, $\alpha = 0$ would imply $\int (f \circ \sigma - f) dv \leq \int (f \circ \sigma - f \pm h) dv$ for all $f, h \in C(\Sigma)$, so that $\int h\, dv = 0$ would contradict $(v, \alpha) \neq (0, 0)$. Besides, for $\alpha < 0$, by taking $-t_1$ and t_2 large enough, one would easily contradict the inequality (2.3).

Define then $\mu := -v/\alpha$. Hence, from (2.3) we obtain $\int g_1\, d\mu - t_1 \geq \int g_2\, d\mu - t_2$ for all $(g_i, t_i) \in C_i$, $i = 1, 2$. As t_1 tends to 0 and t_2 tends to $\kappa + \max(A + g_2)$, we get that

$$\int (g_2 - g_1) d\mu - \max(A + g_2) \leq \kappa \tag{2.4}$$

for any coboundary g_1 and for all continuous function g_2. Therefore, the claim (and the theorem) will be proved when we show that μ is actually a σ-invariant probability.

Notice first that replacing g_2 by cg_2 in inequality (2.4), dividing by $c > 0$, and finally letting c tend to infinity yield $\int g_2 d\mu - \max g_2 \leq 0$. Applying this last inequality to both g_2 and $-g_2$ now leads to

$$\min g_2 \leq \int g_2\, d\mu \leq \max g_2.$$

This shows that μ is a probability measure.

In order to prove that μ is σ-invariant, replace g_1 by cg_1 in (2.4), divide by $c > 0$, and then let c tend to infinity to obtain $\int g_1 d\mu \leq 0$. Since $-g_1$ is also a coboundary, we conclude that $\int g_1 d\mu = 0$ for every coboundary g_1.   □

Chapter 3
Calibrated Sub-actions

We are now concerned with the existence of sub-actions for Lipschitz continuous potentials. In this chapter, with the aid of a suitable operator, we will show that calibrated sub-actions do exist and can be obtained as solutions of a Lax-Oleinik fixed point problem. Instead making use of a version of the classical Schauder-Tychonoff fixed point theorem, we apply a result due to Ishikawa regarding an iteration process for approximating fixed points of nonexpansive mappings, which, at least in theory, opens up interesting simulation possibilities.

3.1   An Iterative Approximation Approach

As we will see along the text, calibrated sub-actions (recall Definition 1.C) have been obtained by different techniques. The existence of calibrated sub-actions by fixed point methods is previously tackled in the concerned literature (see, for instance, [20, 64]). Here we consider the following operator.

Definition 3.A For $A \in C(\Sigma)$, we introduce the associated Lax-Oleinik operator $T_A : C(\Sigma) \to C(\Sigma)$ given by

$$T_A(f)(\mathbf{x}) = \min_{\sigma(\mathbf{y})=\mathbf{x}} [f(\mathbf{y}) - A(\mathbf{y})].$$

A motivating concept for the previous notion is the Lax-Oleinik semigroup, which was employed by Fathi (see [44]) to obtain the so-called weak KAM theorem in the framework of Lagrangian systems. To show the existence of calibrated sub-actions, we will make use of the Lax-Oleinik operator in the context of an adapted version of Ishikawa's theorem as stated below. Its original proof is available in [62, 70]. An interesting proof of the convergence part may be found in [55]. Concerning the vocabulary, by a nonexpansive mapping, we mean a Lipschitz

© The Author(s) 2017   13
E. Garibaldi, *Ergodic Optimization in the Expanding Case*,
SpringerBriefs in Mathematics, DOI 10.1007/978-3-319-66643-3_3

continuous mapping from a subset of a Banach space into itself with Lipschitz
constant less or equal than 1.

A Fixed Point Theorem *Let T be a nonexpansive mapping from a convex and
closed subset Λ of a Banach space into a compact subset of Λ. Then, for any
point ξ_1 of Λ, the sequence $\{\xi_n\}$ defined by $\xi_{n+1} = (\xi_n + T(\xi_n))/2$ converges to
a fixed point of T. Moreover, if diameter$(\Lambda) = \Delta < \infty$, given $\epsilon > 0$, there exists a
computable integer $n_0 = n_0(\epsilon, \Delta) > 0$ (which only depends on ϵ and Δ) such that,
for all nonexpansive mapping T and for all initial point ξ_1, one has $\|\xi_n - T(\xi_n)\| \leq \epsilon$
whenever $n \geq n_0$.*

It is easy to see that the associated Lax-Oleinik operator T_A is Lipschitz continu-
ous, with $\mathrm{Lip}(T_A) \leq 1$. Notice that $T_A(f + c) = T_A(f) + c$ for all $c \in \mathbb{R}$. Therefore,
if we introduce, for a given $[f] \in C(\Sigma)/\mathbb{R}$, the norm $\|[f]\|_\# := \min_{c \in \mathbb{R}} \|f + c\|_\infty$,
we have that T_A induces a well-defined nonexpansive mapping on the Banach space
$(C(\Sigma)/\mathbb{R}, \|\cdot\|_\#)$. Thus, in order to apply the fixed point theorem, we only need the
next lemma. For the constants in the statement, recall the Main Notations in Chap. 1.

Lemma 3.1 *Let A be a Lipschitz continuous potential. Given a constant $M \geq
\frac{\lambda}{1-\lambda} \mathrm{Lip}(S_{K_0}A)$, let Λ_M denote the set of all $[f] \in C(\Sigma)/\mathbb{R}$ such that f is Lipschitz
continuous with $\mathrm{Lip}(f) \leq M$. Then $T_A^{K_0}(\Lambda_M) \subset \Lambda_M$.*

Notice that Λ_M is clearly convex and closed (with respect to topology of the norm
$\|\cdot\|_\#$). As a matter of fact, by the Arzela-Ascoli theorem, Λ_M is compact as a subset
of $(C(\Sigma)/\mathbb{R}, \|\cdot\|_\#)$.

Proof Given $[f] \in \Lambda_M$, we need to argue that $T_A^{K_0}(f)$ is Lipschitz continuous with
Lipschitz constant bounded above by M. For an arbitrary point $\mathbf{x} \in \Sigma$, notice that

$$T_A^{K_0}(f)(\mathbf{x}) = \min_{\sigma^{K_0}(\bar{\mathbf{x}}) = \mathbf{x}} \left[f(\bar{\mathbf{x}}) - S_{K_0}A(\bar{\mathbf{x}}) \right].$$

Let then $\hat{\mathbf{x}} \in \Sigma$ be such that $\sigma^{K_0}(\hat{\mathbf{x}}) = \mathbf{x}$ and $T_A^{K_0}(f)(\mathbf{x}) = f(\hat{\mathbf{x}}) - S_{K_0}A(\hat{\mathbf{x}})$. Given any
point $\mathbf{y} \in \Sigma$, there always exists $\hat{\mathbf{y}} \in \Sigma$ such that $\sigma^{K_0}(\hat{\mathbf{y}}) = \mathbf{y}$ and $d(\hat{\mathbf{x}}, \hat{\mathbf{y}}) \leq \lambda d(\mathbf{x}, \mathbf{y})$.
(Even if $d(\mathbf{x}, \mathbf{y}) = 1$, since all the entries of the matrix $\mathbf{M}^{K_0}$ are strictly positive,
there exists an $\mathbf{M}$-allowed word of length K_0 connecting the first coordinate of $\hat{\mathbf{x}}$ to
the first coordinate of $\mathbf{y}$, so that in this case $\hat{\mathbf{y}}$ is obtained by concatenating this word
with $\mathbf{y}$.) In particular, we have $T_A^{K_0}(f)(\mathbf{y}) \leq f(\hat{\mathbf{y}}) - S_{K_0}A(\hat{\mathbf{y}})$. Notice thus that

$$T_A^{K_0}(f)(\mathbf{y}) - T_A^{K_0}(f)(\mathbf{x}) \leq f(\hat{\mathbf{y}}) - f(\hat{\mathbf{x}}) + S_{K_0}A(\hat{\mathbf{x}}) - S_{K_0}A(\hat{\mathbf{y}})$$

$$\leq \left[\mathrm{Lip}(f) + \mathrm{Lip}(S_{K_0}A) \right] d(\hat{\mathbf{x}}, \hat{\mathbf{y}})$$

$$\leq \left[M + \mathrm{Lip}(S_{K_0}A) \right] \lambda \, d(\mathbf{x}, \mathbf{y}) \leq M d(\mathbf{x}, \mathbf{y}).$$

Since $\mathbf{x}$ and $\mathbf{y}$ play symmetric roles, the proof is complete.   □

We may now state a result that shows the existence of calibrated sub-action by
using an iteration method involving the Lax-Oleinik operator. The importance of
such a methodology was witnessed, for example, in solid-state physics: iterative
algorithm procedure providing calibrated sub-actions for potentials that depend on

a finite number of coordinates (see [47]) was successfully applied in the study of ground states of different models related with the so-called Frenkel-Kontorova model.

Theorem 3.2 *Given a constant $M > 0$, let A be a Lipschitz continuous potential with $Lip(S_{K_0}A) \leq (\lambda^{-1} - 1)M$ and let $f : \Sigma \to \mathbb{R}$ be any Lipschitz continuous function with $Lip(f) \leq M$. Fix an arbitrary point $\mathbf{x}^0 \in \Sigma$. Then the sequence $\{f_n\}$ defined by*

$$f_1 = f - f(\mathbf{x}^0) \quad and$$

$$f_{n+1} = \frac{1}{2}\left(T_A^{K_0}(f_n) - T_A^{K_0}(f_n)(\mathbf{x}^0)\right) + \frac{1}{2}\left(f_n - f_n(\mathbf{x}^0)\right), \quad \forall n \geq 1,$$

converges uniformly to a Lipschitz continuous function v, with $Lip(v) \leq M$, for which

$$u := \min\{v, T_{A-\beta_A}(v), \ldots, T_{A-\beta_A}^{K_0-1}(v)\}$$

is a Lipschitz continuous calibrated sub-action for the potential A. Furthermore, given $\epsilon > 0$, there exists a computable integer $n_0 = n_0(\epsilon, M) > 0$ such that, for all potential A and for all starting function f as before,

$$\left\| f_n - \left(T_A^{K_0}(f_n) - T_A^{K_0}(f_n)(\mathbf{x}^0)\right) \right\|_\infty \leq \epsilon, \quad \forall n \geq n_0.$$

In particular, $\frac{1}{K_0}T_A^{K_0}(f_n)(\mathbf{x}^0) \to \beta_A$ as $n \to \infty$.

Proof Lemma 3.1 allows us to apply the above fixed point theorem to conclude that (for the topology of the norm $\| \cdot \|_\#$) the sequence $\{[f_n]\} \subset \Lambda_M$ converges to a fixed point of $T_A^{K_0}$ that belongs to Λ_M. Let us denote by $[v]$ such a fixed point, where we assume that the Lipschitz continuous function v is exactly the one of its class that vanishes at $\mathbf{x}^0$. Notice that, for all $g \in C(\Sigma)$ with $g(\mathbf{x}^0) = 0$, we have

$$\|[f_n - g]\|_\# = \frac{1}{2}\max_{\mathbf{x},\mathbf{y}\in\Sigma}\left((f_n - g)(\mathbf{x}) - (f_n - g)(\mathbf{y})\right)$$

$$\geq \frac{1}{2}\max_{\mathbf{x}\in\Sigma}(f_n - g)(\mathbf{x}) = \frac{1}{2}\|f_n - g\|_\infty.$$

The above inequality gives us the uniform convergence of $\{f_n\}$ to v, as well as the uniform bounds on the rates for $\|f_n - \left(T_A^{K_0}(f_n) - T_A^{K_0}(f_n)(\mathbf{x}^0)\right)\|_\infty \to 0$.

We show now that $u = \min\{v, \ldots, T_{A-\beta_A}^{K_0-1}(v)\}$ is a calibrated sub-action for the potential A. The fact that $[v]$ is a fixed point of $T_A^{K_0}$ means that there exists a constant $\gamma \in \mathbb{R}$ such that $T_A^{K_0}(v) = v - K_0\gamma$, or equivalently $T_{A-\gamma}^{K_0}(v) = v$. Since Lax-Oleinik operators commute with minima, it is easy to see that

$$T_{A-\gamma}\left(\min\{v, T_{A-\gamma}(v), \ldots, T_{A-\gamma}^{K_0-1}(v)\}\right) = \min\{v, T_{A-\gamma}(v), \ldots, T_{A-\gamma}^{K_0-1}(v)\}.$$

Thus, if we show that $\gamma = \beta_A$, we immediately obtain $T_A(u) = u - \beta_A$, which means that u is a calibrated sub-action. Denote for a moment $\mathfrak{u} := \min\{v, \ldots, T_{A-\gamma}^{K_0-1}(v)\}$. Clearly, we have $A + \mathfrak{u} \circ \sigma - \mathfrak{u} \leq \gamma$ everywhere on Σ. If we take an arbitrary probability $\mu \in \mathcal{M}_\sigma$, then

$$\int A \, d\mu = \int (A + \mathfrak{u} \circ \sigma - \mathfrak{u}) \, d\mu \leq \gamma.$$

Therefore, $\beta_A \leq \gamma$. To show that the equality does hold, fix $\mathbf{y}^0 \in \Sigma$ and, for $k \geq 1$, define inductively a sequence of points $\mathbf{y}^k \in \Sigma$ verifying $\sigma(\mathbf{y}^k) = \mathbf{y}^{k-1}$ and $\mathfrak{u}(\mathbf{y}^{k-1}) = \mathfrak{u}(\mathbf{y}^k) - A(\mathbf{y}^k) + \gamma$. Consider then the sequence of Borel probabilities $\{v_k\}$ given by $v_k := \frac{1}{k} \sum_{j=1}^{k} \delta_{\mathbf{y}^j}$. It is easy to check that any weak* accumulation probability v of $\{v_k\}$ belongs to $\mathcal{M}_\sigma$. We choose one of these accumulation probabilities $v \in \mathcal{M}_\sigma$. Since $\int (A + \mathfrak{u} \circ \sigma - \mathfrak{u}) \, dv_k = \gamma$ for all $k \geq 1$, we get that

$$\int A \, dv = \int (A + \mathfrak{u} \circ \sigma - \mathfrak{u}) \, dv = \gamma.$$

Hence, $\gamma \leq \beta_A$. $\square$

Theorem 3.2 guarantees that there exists at least one Lipschitz continuous calibrated sub-action for A with Lipschitz constant bounded from above by a constant $C(\lambda, K_0, \mathrm{Lip}(A))$. Actually, all calibrated sub-actions for A are Lipschitz continuous with a common upper bound for their Lipschitz constants that depends on the same parameters λ, K_0 and $\mathrm{Lip}(A)$.

Proposition 3.3 *Let* $A : \Sigma \to \mathbb{R}$ *be a Lipschitz continuous potential. If* $u \in C(\Sigma)$ *is a calibrated sub-action for* A, *then* u *is Lipschitz continuous with*

$$Lip(u) \leq \frac{\lambda}{1-\lambda} Lip(S_{K_0}A).$$

Proof Since u is a calibrated sub-action, in particular $u = T_{A-\beta_A}^{K_0}(u)$. Let then $\mathbf{x}^0$ and $\mathbf{y}^0$ be any points of Σ. Define inductively a sequence $\{\mathbf{x}^k\}$ such that, for all $k \geq 0$, $\sigma^{K_0}(\mathbf{x}^{k+1}) = \mathbf{x}^k$ and $u(\mathbf{x}^k) = u(\mathbf{x}^{k+1}) - S_{K_0}(A - \beta_A)(\mathbf{x}^{k+1})$. For $k \geq 1$, we may consider $\mathbf{y}^k \in \Sigma$ such that $\sigma^{K_0}(\mathbf{y}^k) = \mathbf{y}^{k-1}$ and $d(\mathbf{x}^k, \mathbf{y}^k) \leq \lambda d(\mathbf{x}^{k-1}, \mathbf{y}^{k-1})$. Clearly, $u(\mathbf{y}^k) \leq u(\mathbf{y}^{k+1}) - S_{K_0}(A - \beta_A)(\mathbf{y}^{k+1})$. Notice then that

$$u(\mathbf{y}^0) - u(\mathbf{x}^0) \leq u(\mathbf{y}^1) - u(\mathbf{x}^1) + S_{K_0}A(\mathbf{x}^1) - S_{K_0}A(\mathbf{y}^1)$$

$$\vdots$$

$$\leq u(\mathbf{y}^k) - u(\mathbf{x}^k) + \sum_{j=1}^{k} [S_{K_0}A(\mathbf{x}^j) - \acute{S}_{K_0}A(\mathbf{y}^j)].$$

Since $d(\mathbf{x}^k, \mathbf{y}^k) \leq \lambda^k d(\mathbf{x}^0, \mathbf{y}^0) \to 0$ as $k \to \infty$, by passing to the limit, we obtain

$$u(\mathbf{y}^0) - u(\mathbf{x}^0) \le \sum_{j=1}^{\infty} [S_{K_0} A(\mathbf{x}^j) - S_{K_0} A(\mathbf{y}^j)]$$

$$\le \mathrm{Lip}(S_{K_0} A) \sum_{j=1}^{\infty} d(\mathbf{x}^j, \mathbf{y}^j) \le \frac{\lambda}{1-\lambda} \mathrm{Lip}(S_{K_0} A) d(\mathbf{x}^0, \mathbf{y}^0).$$

Since $\mathbf{x}^0$ and $\mathbf{y}^0$ play symmetric roles, the proof is complete.   □

Notice that the previous result shows that the set of continuous calibrated sub-actions for a Lipschitz continuous potential always gives rise to a compact subset of $(C(\Sigma)/\mathbb{R}, \| \cdot \|_\#)$. We will see in Chap. 8 that in general this is not the case when we consider the whole set of continuous sub-actions.

3.2 Asymptotic Behavior of Lax-Oleinik Operators

The iterative procedure described in Theorem 3.2 raises up questions about the behavior of sequences of iterates of a Lax-Oleinik operator. In [90], Nussbaum has shown that, for locally constant potentials, a calibrated sub-action may be always found after a finite number of iterations.

Recall that a function $A : \Sigma \to \mathbb{R}$ is said to be locally constant, or more precisely, to depend on $m + 1$ coordinates if there exists a nonnegative integer m such that $A(\mathbf{x}) = A(\mathbf{y})$ whenever $x_0 = y_0, x_1 = y_1, \dots, x_m = y_m$. Note that in such a case A is Lipschitz continuous.

We highlight the following result due to Nussbaum [90].

Theorem 3.4 *Let $A : \Sigma \to \mathbb{R}$ be a potential that depends on $m + 1$ coordinates. Then, there exists an integer $p > 0$, which is bounded from above by a constant that depends only on the number of $\mathbf{M}$-allowed words of length m, such that, for any function $f : \Sigma \to \mathbb{R}$ depending on m coordinates, the sequence $\{T_{A-\beta_A}^{jp}(f)\}_{j \ge 1}$ is eventually constant. In particular, there exists an integer $j_0 = j_0(f) > 0$ such that the function depending on m coordinates defined by*

$$\min\{T_{A-\beta_A}^{j_0 p}(f), T_{A-\beta_A}^{j_0 p + 1}(f), \dots, T_{A-\beta_A}^{j_0 p + p - 1}(f)\}$$

is a calibrated sub-action for A.

A key ingredient in the proof of this theorem is the finiteness of ω-limit sets of nonexpansive mappings. It seems Weller [102] was the first to notice that, for a nonexpansive mapping of a compact subset of $(\mathbb{R}^M, \| \cdot \|_\infty)$, the ω-limit set of any point (that is, the set of cluster points of its forward orbit) is always finite. Works on cardinality estimation have shown there is an upper bound that depends only on the dimension of the surrounding vector space. For precise estimates, we refer the reader, for example, to [17, 71, 74, 79, 82, 89, 98]. In order to prove Theorem 3.4, we just need a nonexplicit form derived from these results.

A Theorem About ω-Limit Sets of Max-Norm Nonexpansive Mappings *For a nonexpansive mapping $T : C \to C$ of a compact subset C of $(\mathbb{R}^M, \| \cdot \|_\infty)$, the forward orbit of a point of C converges to a periodic orbit whose period is bounded from above by a constant that only depends on M.*

The proof of Theorem 3.4 given here follows essentially the same arguments as the original one [90].

Proof of Theorem 3.4 First recall that a Lax-Oleinik operator is a nonexpansive mapping on $C(\Sigma)$. For a locally constant potential $A(\mathbf{x}) = A(x_0, x_1, \dots, x_m)$, $\mathbf{x} \in \Sigma$, apply Theorem 3.2 to obtain a calibrated sub-action $u \in C(\Sigma)$. Note then that

$$\|T_{A-\beta_A}(g) - u\|_\infty = \|T_{A-\beta_A}(g) - T_{A-\beta_A}(u)\|_\infty \leq \|g - u\|_\infty, \quad \forall\, g \in C(\Sigma),$$

which means that $T_{A-\beta_A}$ maps any closed ball centered at u into itself. It is easy to show that $T_{A-\beta_A}$ preserves the closed subspace of functions that depends on m coordinates. Since these functions are naturally identified with vectors of $\mathbb{R}^M$, where M is the number of **M**-allowed words of length m, we have that $T_{A-\beta_A}$ induces a nonexpansive mapping on the compact subset of $(\mathbb{R}^M, \|\cdot\|_\infty)$ formed from functions g depending on m coordinates such that $\|g - u\|_\infty \leq \|f - u\|_\infty$.

Thanks to the theorem about ω-limit sets of nonexpansive mappings, there exist a function $v : \Sigma \to \mathbb{R}$ that depends on m coordinates and a positive integer p that is bounded by a constant that depends on the number of **M**-allowed words of length m such that

$$T^p_{A-\beta_A}(v) = v \quad \text{and} \quad \lim_{j\to\infty} \|T^{jp}_{A-\beta_A}(f) - v\|_\infty = 0.$$

We claim that $T^{j_0 p}_{A-\beta_A}(f) = v$ for some $j_0 > 0$ (and thus for any other integer greater than j_0). As a matter of fact, $T^p_{A-\beta_A}(v) = v$ implies in particular that $S_p(A + v \circ \sigma - v - \beta_A)(\mathbf{x}) \leq 0$ for all $\mathbf{x} \in \Sigma$. Denoting then $B := A + v \circ \sigma - v - \beta_A$, the equation $T^p_{A-\beta_A}(v) = v$ may be presented as $T^p_B(0) = 0$. Since B depends on $m + 1$ coordinates and $S_p B \leq 0$, there are finitely many values for $S_p B(\mathbf{x}) < 0$, $\mathbf{x} \in \Sigma$, and therefore we define $\eta := \min_{S_p B(\mathbf{x})<0}[-S_p B(\mathbf{x})] > 0$. Let $J > 0$ be an integer such that

$$-\frac{\eta}{2} < T^{Jp}_{A-\beta_A}(f) - v < \frac{\eta}{2}.$$

Since a Lax-Oleinik operator is monotone and commutes with constants and since T^p_B fixes the identically null function, we get

$$-\frac{\eta}{2} < T^{kp}_B\big(T^{Jp}_{A-\beta_A}(f) - v\big) < \frac{\eta}{2}, \quad \forall\, k \geq 1.$$

Given $\mathbf{x} \in \Sigma$, note now that

$$T^{kp}_B\big(T^{Jp}_{A-\beta_A}(f) - v\big)(\mathbf{x}) = \min_{\sigma^{kp}(\bar{\mathbf{x}})=\mathbf{x}} \Big[\big(T^{Jp}_{A-\beta_A}(f) - v\big)(\bar{\mathbf{x}}) - S_{kp}B(\bar{\mathbf{x}})\Big].$$

But if $\bar{\mathbf{x}} \notin (S_{kp}B)^{-1}(0)$, clearly $\left(T_{A-\beta_A}^{Jp}(f) - v\right)(\bar{\mathbf{x}}) - S_{kp}B(\bar{\mathbf{x}}) > -\frac{\eta}{2} + \eta = \frac{\eta}{2}$. Hence, we actually have

$$T_B^{kp}\left(T_{A-\beta_A}^{Jp}(f) - v\right)(\mathbf{x}) = \min_{\substack{\sigma^{kp}(\bar{\mathbf{x}}) = \mathbf{x} \\ \bar{\mathbf{x}} \in (S_{kp}B)^{-1}(0)}} \left(T_{A-\beta_A}^{Jp}(f) - v\right)(\bar{\mathbf{x}}).$$

Since A, f, and v are locally constant, $T_{A-\beta_A}^{Jp}(f) - v$ takes finitely many values on Σ. Therefore, because we are dealing with functions that depend on m coordinates, there must exist $k > j \geq 1$ such that $T_B^{kp}\left(T_{A-\beta_A}^{Jp}(f) - v\right) = T_B^{jp}\left(T_{A-\beta_A}^{Jp}(f) - v\right)$, from which we obtain

$$T_B^{(\ell(k-j)+k)p}\left(T_{A-\beta_A}^{Jp}(f) - v\right) = T_B^{jp}\left(T_{A-\beta_A}^{Jp}(f) - v\right), \quad \forall \ell \geq 0.$$

Since $T_B^p(g - v) = T_{A-\beta_A}^p(g) - v$ for all g, the above equation may be rewritten as

$$T_{A-\beta_A}^{(\ell(k-j)+k+J)p}(f) - v = T_{A-\beta_A}^{(j+J)p}(f) - v, \quad \forall \ell \geq 0.$$

By passing to the limit, we finally see that $T_{A-\beta_A}^{(j+J)p}(f) = v$.

In particular, for $j_0 := J + j$, the fact that

$$\min\{T_{A-\beta_A}^{j_0 p}(f), T_{A-\beta_A}^{j_0 p + 1}(f), \ldots, T_{A-\beta_A}^{j_0 p + p - 1}(f)\} = \min\{v, T_{A-\beta_A}(v), \ldots, T_{A-\beta_A}^{p-1}(v)\}$$

is a calibrated sub-action follows from the periodicity of v and the commutativity between Lax-Oleinik operators and minima.   □

For practical reasons, it may be useful to have in mind that, for a locally constant potential A, efficient algorithms for finding β_A are well known (see, for instance, [68]). Nevertheless, note that in the above proof we have no estimate for j_0, which might be a concern for computational implementations. It is not hard, even so, to better bound the number of consecutive iterates to be considered for determining a calibrated sub-action. We will use the following lemma.

Lemma 3.5 *Let $A : \Sigma \to \mathbb{R}$ be a potential that depends on $m + 1$ coordinates. If M denotes the number of **M**-allowed words of length m, then, for all $k \geq 1$,*

$$\min\{f, T_{A-\beta_A}(f), \ldots, T_{A-\beta_A}^M(f)\} \leq T_{A-\beta_A}^k(f),$$

whenever $f : \Sigma \to \mathbb{R}$ depends on m coordinates.

Proof Without loss of generality, we can pass to a higher block presentation of Σ and assume that A depends on two coordinates. It is then enough to note that, for f a function depending on the first coordinate and $\mathbf{x}$ a point of Σ, there is $j \in \{0, 1, \ldots, r\}$ such that

$$\min_{\sigma^j(\bar{\mathbf{x}}) = \mathbf{x}} \left[f(\bar{x}_0) - \sum_{i=0}^{j-1} \left(A(\bar{x}_i, \bar{x}_{i+1}) - \beta_A\right)\right] \leq \min_{\sigma^k(\bar{\mathbf{x}}) = \mathbf{x}} \left[f(\bar{x}_0) - \sum_{i=0}^{k-1} \left(A(\bar{x}_i, \bar{x}_{i+1}) - \beta_A\right)\right].$$

Indeed, for $k > r$, a subword of $(\bar{x}_0, \bar{x}_1, \ldots, \bar{x}_{k-1})$ with identical endpoints $\bar{x}_\iota = \bar{x}_{\iota'}$, $0 \leq \iota < \iota' \leq k - 1$, defines by repetition a periodic point $\bar{\bar{x}} \in \Sigma$. From (2.2), we thus have

$$\sum_{i=\iota}^{\iota'-1} \left(A(\bar{x}_i, \bar{x}_{i+1}) - \beta_A \right) = \sum_{\bar{\bar{y}} \in \mathrm{orb}(\bar{\bar{x}})} (A - \beta_A)(\bar{\bar{y}}) \leq 0.$$

The claimed inequality follows from successive applications of this remark. □
   We have thus an immediate consequence.

Proposition 3.6 *Let $A : \Sigma \to \mathbb{R}$ be a potential that depends on $m + 1$ coordinates. Denote by M the number of $\mathbf{M}$-allowed words of length m. Given any $f : \Sigma \to \mathbb{R}$ depending on m coordinates, there is an integer $J_0 = J_0(f) > 0$ such that*

$$\min \left\{ T_{A-\beta_A}^{J_0}(f), T_{A-\beta_A}^{J_0+1}(f), \ldots, T_{A-\beta_A}^{J_0+M}(f) \right\}$$

is a calibrated sub-action for A that depends on m coordinates.

Proof Let $g : \Sigma \to \mathbb{R}$ be a function depending on m coordinates such that $g \leq T_{A-\beta_A}(g)$. From the conclusion of Theorem 3.4 applied to g, thanks to the monotonicity of Lax-Oleinink operators, we obtain that the sequence $\{T_{A-\beta_A}^j(g)\}_{j \geq j_0 p}$ is constant and therefore $T_{A-\beta_A}^{j_0 p}(g)$ is a calibrated sub-action. Lemma 3.5 allows us to apply this observation to $g := \min\{f, T_{A-\beta_A}(f), \ldots, T_{A-\beta_A}^M(f)\}$.   □

Chapter 4
Aubry Set

We begin now to develop a variational methodology for ergodic optimization, in which orbits will play an increasing role as optimizing ingredients. The first notion we present is the Aubry set, the part of the non-wandering set that characterizes the maximizing probabilities. Roughly speaking, the Aubry set is formed by the non-wandering points whose orbits have maximal Birkhoff sums. In the sequel, we give the precise definition and some examples for this key concept.

4.1 A Maximizing Non-wandering Set

We discuss in this chapter basic properties of the Aubry set, a notion first introduced in ergodic optimization by Contreras, Lopes and Thieullen [40]. Initially, we remark that the set equality (1.5) is in fact a proposition to be proved, since the usual definition of the Aubry set is as follows.

Definition 4.A Given a potential $A \in C(\Sigma)$, we call $\mathbf{x} \in \Sigma$ an Aubry point when, for all $\epsilon > 0$, there exist a point $\mathbf{y} \in \Sigma$ and an integer $n > 0$ such that

$$d(\mathbf{x}, \mathbf{y}) < \epsilon, \ d(\mathbf{x}, \sigma^n(\mathbf{y})) < \epsilon \text{ and } |S_n(A - \beta_A)(\mathbf{y})| < \epsilon. \tag{4.1}$$

We call Aubry set and denote by $\Omega(A)$ the collection of all Aubry points.

By arguing that it contains the limits of convergent sequences of Aubry points, it is easy to show that the Aubry set is always closed and thus compact. Note that $\Omega(A + f \circ \sigma - f - c) = \Omega(A)$ for every function $f \in C(\Sigma)$ and for any constant $c \in \mathbb{R}$. In particular, for $f = A$ and $c = 0$, we see that $\Omega(A \circ \sigma) = \Omega(A)$, from which we may obtain the σ-invariance of the Aubry set. In fact, if $\mathbf{x} \in \Omega(A \circ \sigma)$, then, for all integer $j > 0$, there are $\mathbf{y} \in \Sigma$ and $n > 0$ such that

$$d(\mathbf{x}, \mathbf{y}) < \lambda^j, \ d(\mathbf{x}, \sigma^n(\mathbf{y})) < \lambda^j \text{ and } |S_n(A \circ \sigma - \beta_A)(\mathbf{y})| < \lambda^j.$$

© The Author(s) 2017
E. Garibaldi, *Ergodic Optimization in the Expanding Case*,
SpringerBriefs in Mathematics, DOI 10.1007/978-3-319-66643-3_4

Since $d(\sigma(\mathbf{x}), \sigma(\mathbf{y})) < \lambda^{j-1}$, $d(\sigma(\mathbf{x}), \sigma^n(\sigma(\mathbf{y}))) < \lambda^{j-1}$ and $S_n(A \circ \sigma - \beta_A)(\mathbf{y}) = S_n(A - \beta_A)(\sigma(\mathbf{y}))$, we have shown that $\sigma(\mathbf{x}) \in \Omega(A)$. Therefore, $\sigma(\Omega(A)) \subset \Omega(A)$.

Furthermore, if the existence of a sub-action $u \in C(\Sigma)$ is guaranteed, then we have

$$\Omega(A) \subset (A + u \circ \sigma - u)^{-1}(\beta_A). \tag{4.2}$$

Indeed, it is enough to show that $\Omega(A + u \circ \sigma - u) \subset (A + u \circ \sigma - u)^{-1}(\beta_A)$. Given $x \in \Omega(A + u \circ \sigma - u)$, for any positive integer j, there exist $\mathbf{y}^j \in \Sigma$ and $n_j > 0$ such that

$$d(\mathbf{x}, \mathbf{y}^j) < \frac{1}{j}, \ \ d(\mathbf{x}, \sigma^{n_j}(\mathbf{y}^j)) < \frac{1}{j} \ \ \text{and} \ \ \left| S_{n_j}(A + u \circ \sigma - u - \beta_A)(\mathbf{y}^j) \right| < \frac{1}{j}.$$

Since u is a sub-action, note that

$$-\frac{1}{j} \le S_{n_j}(A + u \circ \sigma - u - \beta_A)(\mathbf{y}^j) \le (A + u \circ \sigma - u - \beta_A)(\mathbf{y}^j) \le 0.$$

Therefore, by passing to the limit, we get $A(\mathbf{x}) + u \circ \sigma(\mathbf{x}) - u(\mathbf{x}) - \beta_A = 0$. In other terms, $\mathbf{x} \in \Omega(A + u \circ \sigma - u) = \Omega(A)$ implies $\mathbf{x} \in (A + u \circ \sigma - u)^{-1}(\beta_A)$.

Whenever the existence of a maximizing probability may be guaranteed, as a consequence of Atkinson's characterization of recurrence of random walks [7], the corresponding Aubry set is nonempty. In particular, this is true for any continuous potential over any continuous transformation of a compact metric space. Since these notes are intended to be as self-contained as possible and suitable for a wider range of readers, let us now present a sufficient condition for the Aubry set to be nonempty.

We will need the following notion: if $u \in C(\Sigma)$ is an arbitrary sub-action for a continuous potential A, we say that a sequence $\{\mathbf{x}^k\} \subset \Sigma$ is a u-optimal trajectory if $\sigma(\mathbf{x}^k) = \mathbf{x}^{k-1}$ and $A(\mathbf{x}^k) + u(\mathbf{x}^{k-1}) - u(\mathbf{x}^k) = \beta_A$ for each k. We have proved in the previous chapter that there exists a calibrated sub-action $u \in C(\Sigma)$ if the potential $A : \Sigma \to \mathbb{R}$ is Lipschitz continuous. Therefore, in this case, for every point $\mathbf{x} \in \Sigma$, one can evidently find a u-optimal trajectory $\{\mathbf{x}^k\}_{k \ge 0}$ with $\mathbf{x}^0 = \mathbf{x}$.

The next proposition ensures the existence of Aubry points for continuous potentials that admit continuous sub-actions.

Proposition 4.1 *Let $A : \Sigma \to \mathbb{R}$ be a continuous potential. Suppose the function $u \in C(\Sigma)$ is an arbitrary sub-action for A. Then, there exist u-optimal trajectories and any accumulation point of a u-optimal trajectory belongs to $\Omega(A)$.*

Proof Let $\mu \in \mathcal{M}_\sigma$ be A-maximizing. For every point $\mathbf{x} \in \text{supp}(\mu)$, we use the fact that $\sigma(\text{supp}(\mu)) = \text{supp}(\mu)$ and the characterization (1.4) to construct a u-optimal trajectory $\{\mathbf{x}^k\}_{k \ge 0} \subset \text{supp}(\mu)$ with $\mathbf{x}^0 = \mathbf{x}$.

Consider now any u-optimal trajectory $\{\mathbf{x}^k\} \subset \Sigma$. Let $\bar{\mathbf{x}}$ be the limit of a subsequence $\{\mathbf{x}^{k_j}\}$. For $\epsilon > 0$, let $\eta \in (0, 2\epsilon)$ be such that, if $\mathbf{x}, \mathbf{y} \in \Sigma$ satisfy $d(\mathbf{x}, \mathbf{y}) < \eta$, then $|u(\mathbf{x}) - u(\mathbf{y})| < \epsilon$. There exists an integer $J > 0$ for which

$d(\mathbf{x}^{k_j}, \bar{\mathbf{x}}) < \eta/2$ whenever $j \geq J$. Set thus $\mathbf{y} := \mathbf{x}^{k_J+1}$ and $n := k_{J+1} - k_J$. Note that $d(\mathbf{x}, \mathbf{y}) = d(\mathbf{x}, \mathbf{x}^{k_J+1}) < \eta/2 < \epsilon$ and $d(\mathbf{x}, \sigma^n(\mathbf{y})) = d(\mathbf{x}, \mathbf{x}^{k_J}) < \eta/2 < \epsilon$. Moreover, since

$$|S_n(A - \beta_A)(\mathbf{y})| = \left| \sum_{j=0}^{n-1} [u \circ \sigma^j(\mathbf{y}) - u \circ \sigma^{j+1}(\mathbf{y})] \right|$$

$$= |u(\mathbf{y}) - u(\sigma^n(\mathbf{y}))| = |u(\mathbf{x}^{k_J+1}) - u(\mathbf{x}^{k_J})|,$$

from $d(\mathbf{x}^{k_J+1}, \mathbf{x}^{k_J}) < \eta$ we get $|S_n(A - \beta_A)(\mathbf{y})| < \epsilon$. Hence, $\bar{\mathbf{x}} \in \Omega(A)$. $\qquad \square$

Assume that $A : \Sigma \to \mathbb{R}$ is a continuous potential that admits a continuous sub-action. As the corresponding Aubry set is nonempty, compact, and invariant, there exists at least one σ-invariant Borel probability measure μ whose support lies on $\Omega(A)$. Thanks to the inclusion (4.2), one concludes immediately from (1.4) that such a measure μ is an A-maximizing probability. In other terms, any invariant probability whose support is contained in the Aubry set is a maximizing measure.

The converse also holds, that is, every A-maximizing measure must be supported on $\Omega(A)$. As remarked in Chap. 1, this description of m_A may be seen as a direct consequence of the set equality (1.5), which remains to be proved. In Chap. 7, for a Lipschitz continuous potential A, we will show the existence of a continuous sub-action u such that $\Omega(A) = (A + u \circ \sigma - u)^{-1}(\beta_A)$. Such a function is called a separating sub-action. Notice that the existence of a separating sub-action guarantees at once the set equality (1.5) and hence

$$m_A = \{\mu \in \mathcal{M}_\sigma : \text{supp}(\mu) \subset \Omega(A)\}.$$

We remark that customarily (see, for instance [40, 75]) such a property is obtained as a direct consequence of Atkinson's theorem [7]. Therefore, the existence of separating sub-actions provides an alternative approach for describing maximizing probabilities as those invariant measures whose support lies on the Aubry set.

In some cases, it is very easy to characterize the Aubry points. For instance, let $\mu \in \mathcal{M}_\sigma$ be uniquely ergodic. If $A(\mathbf{x}) = -d(\mathbf{x}, \text{supp}(\mu))$, then $\Omega(A) = \text{supp}(\mu)$. As a matter of fact, since the identically constant null function is a sub-action for A and $\beta_A = 0$, from (4.2) we have the inclusion $\Omega(A) \subset A^{-1}(0) = \text{supp}(\mu)$, which by minimality must be an equality. More generally, if $\mu_i \in \mathcal{M}_\sigma$, $i = 1, \ldots, k$, for the Lipschitz continuous potential $A(\mathbf{x}) = -\prod_{i=1}^k d(\mathbf{x}, \text{supp}(\mu_i))$, we realize that $\Omega(A) = \cup_{i=1}^k \text{supp}(\mu_i)$. Thanks to Poincaré's recurrence theorem, this is a consequence of the next result.

Proposition 4.2 *For $i = 1, \ldots, k$, suppose that $\Omega_i \subset \Sigma$ is a nonempty compact invariant set whose non-wandering points form a dense subset. The Aubry set of the Lipschitz continuous potential $A(\mathbf{x}) := -\prod_{i=1}^k d(\mathbf{x}, \Omega_i)$ is then $\Omega(A) = \cup_{i=1}^k \Omega_i$.*

Proof Obviously $A \leq 0 = \beta_A$. If $k = 1$, from (4.2) we see that $\Omega(A) \subset A^{-1}(0) = \Omega_1$. Besides, any point $\mathbf{y} \in \Omega_1$ clearly verifies $S_n A(\mathbf{y}) = 0$ for all $n \geq 0$. Since the non-wandering points of Ω_1 are dense, given any $\mathbf{x} \in \Omega_1$ and $\epsilon > 0$, there exist

$y \in \Omega_1$ and $n > 0$ such that $d(\mathbf{x}, \mathbf{y}) < \epsilon$ and $d(\mathbf{x}, \sigma^n(\mathbf{y})) < \epsilon$. Thus, $\mathbf{x}$ is an Aubry point and $\Omega_1 \subset \Omega(A)$. We have shown that the set equality holds for $k = 1$.

For the general case, we remark that $\hat{B} \leq B \leq 0$ and $\beta_{\hat{B}} = 0$ imply $\Omega(\hat{B}) \subset \Omega(B)$. Indeed, this follows immediately from the fact the $\beta_B = 0$ and from the trivial inequalities $S_n\hat{B} \leq S_nB \leq 0$ for all $n \geq 0$.

Note now that, if we introduce $\hat{A}(\mathbf{x}) := -d(\mathbf{x}, \cup_{i=1}^k \Omega_i)$, then $\hat{A} \leq A \leq 0$ and $\beta_{\hat{A}} = 0$. From the particular case and the previous remark, we get $\cup_{i=1}^k \Omega_i = \Omega(\hat{A}) \subset \Omega(A)$. But we can apply (4.2) to obtain $\Omega(A) \subset A^{-1}(0) = \cup_{i=1}^k \Omega_i$. The proof is complete.   □

A well-known characterization of the Aubry set is obtained for locally constant potentials.

Theorem 4.3 *Let A be a locally constant potential on a topologically mixing subshift of finite type. Then $\Omega(A)$ is a subshift of finite type.*

Proof By passing to a higher block presentation of Σ, we may assume without loss of generality that A depends just on two coordinates. If v is a locally constant sub-action whose existence was guaranteed by Theorem 3.4, replacing $A(x_0, x_1)$ by $A(x_0, x_1) + v(x_1) - v(x_0) - \beta_A$, we suppose from now on that $A \leq 0$ and $\beta_A = 0$. Consider then the $r \times r$ transition matrix $\mathbf{N}$ defined by $\mathbf{N}(i, j) = 1$ if, and only if, there exists an $\mathbf{M}$-allowed word $(w_0, w_1, \ldots, w_n)$, with $w_0 = w_n = i$ and $w_1 = j$, such that $A(w_0, w_1) + \cdots + A(w_{n-1}, w_n) = 0$.

Claim $\Omega(A) = \Sigma_A := \{\mathbf{x} \in \{1, \ldots, r\}^{\mathbb{N}} : \mathbf{N}(x_j, x_{j+1}) = 1 \text{ for all } j \geq 0\}$.

Indeed, given $\mathbf{x} = (x_0, x_1, \ldots) \in \Omega(A)$, since $\sigma(\Omega(A)) \subset \Omega(A)$, in order to show that $\Omega(A) \subset \Sigma_A$, we only need to argue that $\mathbf{N}(x_0, x_1) = 1$. Denote then $\eta = \min_{A(i,j) < 0}[-A(i,j)]$ and notice that there exist a point $\mathbf{y} \in \Sigma$ and a positive integer n such that

$$d(\mathbf{x}, \mathbf{y}) < \lambda, \quad d(\mathbf{x}, \sigma^n(\mathbf{y})) < \lambda \text{ and } -\eta < \sum_{k=0}^{n-1} A \circ \sigma^k(\mathbf{y}) \leq 0.$$

Therefore, $(y_0, y_1, \ldots, y_n)$ is an $\mathbf{M}$-allowed word such that $y_0 = y_n = x_0$, $y_1 = x_1$, and, by the definition of η, $A(y_0, y_1) + \cdots + A(y_{n-1}, y_n) = 0$.

Conversely, for $\mathbf{x} \in \Sigma_A$, given a positive integer m and $l \in \{0, \ldots, m\}$, there exists an $\mathbf{M}$-allowed word $(w_0^l, w_1^l, \ldots, w_{n_l}^l)$, with $w_0^l = w_{n_l}^l = x_l$ and $w_1^l = x_{l+1}$, such that $A(w_0^l, w_1^l) + \cdots + A(w_{n_l-1}^l, w_{n_l}^l) = 0$. Writing concisely $w^l := (w_1^l, \ldots, w_{n_l-1}^l)$, it is easy to see that the concatenation $(x_0, \ldots, x_m)w^m w^{m-1} \cdots w^0$ defines an $\mathbf{M}$-allowed word. Moreover, by repetition this word gives us a periodic point $\mathbf{y} \in \Sigma$, with period $n = n_0 + n_1 + \cdots + n_m$. Clearly, we have $d(\mathbf{x}, \mathbf{y}) = d(\mathbf{x}, \sigma^n(\mathbf{y})) < \lambda^m$. Since

$$\sum_{k=0}^{n-1} A \circ \sigma^k(\mathbf{y}) = \sum_{l=0}^{m} \sum_{k=0}^{n_l-1} A(w_k^l, w_{k+1}^l) = 0,$$

we have shown that $\mathbf{x} \in \Omega(A)$.   □

We point out that, as consequence of the argument that we have presented in the previous proof, we also obtained

$$\Omega(A) = \overline{\{\operatorname{supp}(\mu) \mid \mu A\text{-maximizing periodic probability}\}}$$

for a locally constant potential A. In general, there is a very simple criterion for determining the existence of a periodic Aubry point.

Proposition 4.4 *Let $A : \Sigma \to \mathbb{R}$ be a continuous potential. Suppose that, for some $x \in \Omega(A)$, one can find a bounded family of positive integers $\{n(\epsilon)\}_{\epsilon>0}$ (and an associated family of points $\{y(\epsilon)\}_{\epsilon>0}$) for which (4.1) holds for every $\epsilon > 0$. Then x is a periodic point.*

Proof By considering an accumulation point $y \in \Sigma$ as ϵ tends to 0, for some integer $N > 0$ the passage to the limit evidently yields $d(x, y) = 0$ and $d(x, \sigma^N(y)) = 0$, that is, $y = x = \sigma^N(y)$. □

Hence, when $x \in \Omega(A)$ is periodic, we can (artificially) take a sufficiently large multiple of its period in order to ensure the following result.

Corollary 4.5 *For $A \in C(\Sigma)$, x is an Aubry point if, and only if, for all $\epsilon > 0$ and $\ell > 0$, there exist a point $y \in \Sigma$ and an integer $n > \ell$ such that $d(x, y) < \epsilon$, $d(x, \sigma^n(y)) < \epsilon$ and $|S_n(A - \beta_A)(y)| < \epsilon$.*

Proof If $x \in \Omega(A)$ is not periodic, we apply the previous proposition. Besides, for $x \in \Omega(A)$ a periodic point of period m, as we have already remarked, the fact that the associated periodic probability has its support (that is, the orbit of x) contained in $\Omega(A)$ implies that this measure is A-maximizing, so that $S_{km}A(x) = km\beta_A$ for all $k \geq 1$. Therefore, in this case, we consider $y = x$ and $n = km > \ell$. □

Chapter 5
Mañé Potential and Peierls Barrier

We will embark upon the task of interpreting discrete orbits as parameterized curves in such a way that an important machinery involving action functionals will be available in ergodic optimization theory. The concepts that will be discussed in this chapter, namely, the Peierls barrier and the Mañé potential go back to the contributions of both Mather and Mañé in Lagrangian systems. We concentrate our attention here on their essential properties, highlighting relevant similarities to their predecessors as well as peculiarities arising from the discrete-dynamics scenario. We also take the opportunity to shed light on an interconnection between ergodic optimization and max algebra, by showing how the Mañé potential is related to the Kleene star.

5.1 Action Functionals in Ergodic Optimization

The Mañé potential and the Peierls barrier have a key role in the study of sub-actions and therefore of the optimizing probabilities. We introduce in the next definition these two notions of action potential between two points.

Definition 5.A Let $A \in C(\Sigma)$.

i. We call Mañé potential (associated with A) the function ϕ_A defined on $\Sigma \times \Sigma$ by

$$\phi_A(\mathbf{x}, \mathbf{y}) := \lim_{\epsilon \to 0} \inf_{n>0} \inf_{\substack{d(\mathbf{z},\mathbf{x})<\epsilon \\ d(\sigma^n(\mathbf{z}),\mathbf{y})<\epsilon}} [-S_n(A - \beta_A)(\mathbf{z})].$$

ii. We call Peierls barrier (associated with A) the function h_A defined on $\Sigma \times \Sigma$ by

$$h_A(\mathbf{x}, \mathbf{y}) := \lim_{\epsilon \to 0} \liminf_{n \to \infty} \inf_{\substack{d(\mathbf{z},\mathbf{x})<\epsilon \\ d(\sigma^n(\mathbf{z}),\mathbf{y})<\epsilon}} [-S_n(A - \beta_A)(\mathbf{z})].$$

© The Author(s) 2017
E. Garibaldi, *Ergodic Optimization in the Expanding Case*,
SpringerBriefs in Mathematics, DOI 10.1007/978-3-319-66643-3_5

Obviously $\phi_A \leq h_A$. Moreover, it is not difficult to see that both functions are lower semi-continuous. The next lemma will be useful for the discussion of further properties of the Mañé potential and the Peierls barrier. When convenient, we will adopt from now on the concise notation $a \wedge b := \min\{a, b\}$.

Lemma 5.1 *Let $A : \Sigma \to \mathbb{R}$ be a Lipschitz continuous potential. Given $\epsilon > 0$, $\epsilon' \geq 0$ and $l \geq 0$, consider the auxiliary function*

$$\mathfrak{S}_l^{\epsilon,\epsilon'}(\mathbf{x}, \mathbf{y}) := \inf_{n \geq l} \inf_{\substack{d(\mathbf{w},\mathbf{x}) \leq \epsilon \\ d(\sigma^n(\mathbf{w}),\mathbf{y}) \leq \epsilon'}} [-S_n(A - \beta_A)(\mathbf{w})], \qquad \forall\, \mathbf{x}, \mathbf{y} \in \Sigma. \tag{5.1}$$

Then, for all integers $k, k', \bar{k} > 0$, $l, \bar{l} \geq 0$, for any constant $\bar{\epsilon}' \geq 0$, and for all points $\mathbf{x}, \mathbf{y}, \mathbf{z} \in \Sigma$,

$$\mathfrak{S}_{l+\bar{l}}^{\lambda^{k \wedge (l+k' \wedge \bar{k})}, \bar{\epsilon}'}(\mathbf{x}, \mathbf{z}) \leq \mathfrak{S}_l^{\lambda^k, \lambda^{k'}}(\mathbf{x}, \mathbf{y}) + \mathfrak{S}_{\bar{l}}^{\lambda^{\bar{k}}, \bar{\epsilon}'}(\mathbf{y}, \mathbf{z}) + \frac{Lip(A)}{1 - \lambda} \lambda^{k' \wedge \bar{k}}. \tag{5.2}$$

Notice that we clearly have

$$\phi_A(\mathbf{x}, \mathbf{y}) = \lim_{\epsilon \to 0} \mathfrak{S}_1^{\epsilon,\epsilon}(\mathbf{x}, \mathbf{y}) \qquad \text{and} \qquad h_A(\mathbf{x}, \mathbf{y}) = \limsup_{\epsilon \to 0} \mathfrak{S}_l^{\epsilon,\epsilon}(\mathbf{x}, \mathbf{y}).$$

Proof Fix a constant $\eta > 0$. Let $\mathbf{x}, \mathbf{y}, \mathbf{z} \in \Sigma$. Consider integers $k, k', \bar{k} > 0$, $l, \bar{l} \geq 0$ and a constant $\bar{\epsilon}' \geq 0$. There exist a point $\mathbf{w} = (w_0, w_1, \ldots) \in \Sigma$ and an integer $n \geq l$ such that $d(\mathbf{w}, \mathbf{x}) \leq \lambda^k$, $d(\sigma^n(\mathbf{w}), \mathbf{y}) \leq \lambda^{k'}$ and

$$\mathfrak{S}_l^{\lambda^k, \lambda^{k'}}(\mathbf{x}, \mathbf{y}) + \eta > -S_n(A - \beta_A)(\mathbf{w}).$$

There also exist a point $\overline{\mathbf{w}} = (\overline{w}_0, \overline{w}_1, \ldots) \in \Sigma$ and an integer $\bar{n} \geq \bar{l}$ such that $d(\overline{\mathbf{w}}, \mathbf{y}) \leq \lambda^{\bar{k}}$, $d(\sigma^{\bar{n}}(\overline{\mathbf{w}}), \mathbf{z}) \leq \bar{\epsilon}'$ and

$$\mathfrak{S}_{\bar{l}}^{\lambda^{\bar{k}}, \bar{\epsilon}'}(\mathbf{y}, \mathbf{z}) + \eta > -S_{\bar{n}}(A - \beta_A)(\overline{\mathbf{w}}).$$

Define $\overline{\overline{\mathbf{w}}} := (w_0, w_1, \ldots, w_n, \overline{w}_1, \overline{w}_2, \ldots) \in \Sigma$. By construction, we have that $(y_0, \ldots, y_{k'}) = (w_n, \ldots, w_{n+k'})$ and $(y_0, \ldots, y_{\bar{k}}) = (\overline{w}_0, \ldots, \overline{w}_{\bar{k}})$. So analyzing the relative positions of k and $n + k' \wedge \bar{k}$, it is easy to see that $d(\overline{\overline{\mathbf{w}}}, \mathbf{x}) \leq \lambda^{k \wedge (n+k' \wedge \bar{k})}$. Besides, obviously $d(\sigma^{n+\bar{n}}(\overline{\overline{\mathbf{w}}}), \mathbf{z}) = d(\sigma^{\bar{n}}(\overline{\mathbf{w}}), \mathbf{z}) \leq \bar{\epsilon}'$. Therefore, we obtain

$$\mathfrak{S}_{l+\bar{l}}^{\lambda^{k \wedge (n+k' \wedge \bar{k})}, \bar{\epsilon}'}(\mathbf{x}, \mathbf{z}) \leq -S_{n+\bar{n}}(A - \beta_A)(\overline{\overline{\mathbf{w}}})$$

$$= -S_n(A - \beta_A)(\overline{\overline{\mathbf{w}}}) - S_{\bar{n}}(A - \beta_A)(\overline{\mathbf{w}}).$$

Notice now that

$$-S_n(A - \beta_A)(\overline{\overline{\mathbf{w}}}) < \mathfrak{S}_l^{\lambda^k, \lambda^{k'}}(\mathbf{x}, \mathbf{y}) + \eta + S_n A(\mathbf{w}) - S_n A(\overline{\overline{\mathbf{w}}})$$

$$\leq \mathfrak{S}_l^{\lambda^k, \lambda^{k'}}(\mathbf{x}, \mathbf{y}) + \eta + \mathrm{Lip}(A)\big(\lambda^{n+k' \wedge \tilde{k}} + \cdots + \lambda^{k' \wedge \tilde{k}}\big).$$

Thus, considering $\eta > 0$ arbitrarily small, we get

$$\mathfrak{S}_{l+\tilde{l}}^{\lambda^{k \wedge (l + k' \wedge \tilde{k})}, \tilde{\epsilon}'}(\mathbf{x}, \mathbf{z}) \leq \mathfrak{S}_l^{\lambda^k, \lambda^{k'}}(\mathbf{x}, \mathbf{y}) + \mathfrak{S}_{\tilde{l}}^{\lambda^{\tilde{k}}, \tilde{\epsilon}'}(\mathbf{y}, \mathbf{z}) + \frac{\mathrm{Lip}(A)}{1 - \lambda} \lambda^{k' \wedge \tilde{k}}.$$

□

The following proposition summarizes the main properties of both action potentials.

Proposition 5.2 *Assume the potential* $A : \Sigma \to \mathbb{R}$ *is Lipschitz continuous. Then*

  i. *for any sub-action* $u \in C(\Sigma)$, $\phi_A(\mathbf{x}, \mathbf{y}) \geq u(\mathbf{y}) - u(\mathbf{x})$;
 ii. *for any points* $\mathbf{x}, \mathbf{y}, \mathbf{z} \in \Sigma$,

$$\phi_A(\mathbf{x}, \mathbf{z}) \leq \phi_A(\mathbf{x}, \mathbf{y}) + \phi_A(\mathbf{y}, \mathbf{z}), \tag{5.3}$$

$$h_A(\mathbf{x}, \mathbf{z}) \leq \phi_A(\mathbf{x}, \mathbf{y}) + h_A(\mathbf{y}, \mathbf{z}), \tag{5.4}$$

$$h_A(\mathbf{x}, \mathbf{z}) \leq h_A(\mathbf{x}, \mathbf{y}) + h_A(\mathbf{y}, \mathbf{z}); \tag{5.5}$$

iii. $\mathbf{x} \in \Omega(A) \Leftrightarrow \phi_A(\mathbf{x}, \mathbf{x}) = 0 \Leftrightarrow h_A(\mathbf{x}, \mathbf{x}) = 0$;
 iv. *if* $\mathbf{x} \in \Omega(A)$, *then* $\phi_A(\mathbf{x}, \cdot) = h_A(\mathbf{x}, \cdot)$ *is a Lipschitz continuous calibrated sub-action.*

We highlight that the Mañé potential or the Peierls barrier allows exhibiting a calibrated sub-action without using some kind of Lax-Oleinik fixed point method. Notice also that, since ϕ_A is lower semi-continuous, the inequality (5.4) is equivalent to

$$h_A(\mathbf{x}, \mathbf{z}) \leq \liminf_{k \to \infty} \phi_A(\mathbf{x}, \mathbf{y}^k) + h_A(\mathbf{y}, \mathbf{z})$$

for any points $\mathbf{x}, \mathbf{y}, \mathbf{z} \in \Sigma$ and any sequence $\{\mathbf{y}^k\}$ converging to $\mathbf{y}$.

Proof
Item i. Let $u \in C(\Sigma)$ be any sub-action for A. Given $\eta > 0$, there exists $\epsilon > 0$ such that, if $\bar{\mathbf{x}}, \bar{\mathbf{y}} \in \Sigma$ satisfy $d(\bar{\mathbf{x}}, \bar{\mathbf{y}}) < \epsilon$, then $|u(\bar{\mathbf{x}}) - u(\bar{\mathbf{y}})| < \eta/2$. Therefore, for $\mathbf{z} \in \Sigma$ with $d(\mathbf{z}, \mathbf{x}) < \epsilon$ and $d(\sigma^n(\mathbf{z}), \mathbf{y}) < \epsilon$, we clearly have

$$u(\mathbf{y}) - u(\mathbf{x}) - \eta < u(\sigma(\mathbf{z})) - u(\mathbf{z}) \leq -S_n(A - \beta_A)(\mathbf{z}).$$

Hence, we deduce that $u(\mathbf{y}) - u(\mathbf{x}) - \eta \leq \phi_A(\mathbf{x}, \mathbf{y})$ and the claimed inequality follows by considering $\eta > 0$ arbitrarily small.

Item ii. Our strategy consists in showing all the inequalities as consequences of the same argumentation. Notice then that, for $k = k' = \bar{k}$ and $\bar{\epsilon}' = \lambda^k$, a particular case of inequality (5.2) is

$$\mathfrak{S}_{l+\bar{l}}^{\lambda^k, \lambda^k}(\mathbf{x}, \mathbf{z}) \leq \mathfrak{S}_l^{\lambda^k, \lambda^k}(\mathbf{x}, \mathbf{y}) + \mathfrak{S}_{\bar{l}}^{\lambda^k, \lambda^k}(\mathbf{y}, \mathbf{z}) + \frac{\mathrm{Lip}(A)}{1 - \lambda} \lambda^k. \tag{5.6}$$

All the inequalities we want to prove follow from the former one. Indeed, if we set $l = 0$ and $\bar{l} = 1$, since $\mathfrak{S}_0^{\lambda^k, \lambda^k}(\mathbf{x}, \mathbf{y}) \leq \mathfrak{S}_1^{\lambda^k, \lambda^k}(\mathbf{x}, \mathbf{y})$, by taking the limit when k tends to infinity, we deduce inequality (5.3). In order to guarantee (5.4), after fixing $l = 1$ in last inequality, we just need to take the supremum over $\bar{l}$ before passing to the limit. At last, we initially consider the supremum over l and then repeat the previous process to obtain inequality (5.5).

Item iii. It follows from item i that $\phi_A(\mathbf{x}, \mathbf{x}) \geq 0$. Since $h_A \geq \phi_A$, obviously $h_A(\mathbf{x}, \mathbf{x}) = 0$ implies $\phi_A(\mathbf{x}, \mathbf{x}) = 0$.

We first show that, if $\phi_A(\mathbf{x}, \mathbf{x}) = 0$, then $\mathbf{x} \in \Omega(A)$. Let $u \in C(\Sigma)$ be a sub-action for A. Given $\epsilon > 0$, there exists $\eta \in (0, \epsilon)$ such that, if $\bar{\mathbf{x}}, \bar{\mathbf{y}} \in \Sigma$ satisfy $d(\bar{\mathbf{x}}, \bar{\mathbf{y}}) < 2\eta$, then $|u(\bar{\mathbf{x}}) - u(\bar{\mathbf{y}})| < \epsilon$. Notice that $\mathfrak{S}_1^{\eta, \eta}(\mathbf{x}, \mathbf{x})$ increases to $\phi_A(\mathbf{x}, \mathbf{x}) = 0$ as η tends to 0. In particular, one can find a point $\mathbf{y} \in \Sigma$ and an integer $n > 0$ such that $d(\mathbf{x}, \mathbf{y}) < \eta$, $d(\mathbf{x}, \sigma^n(\mathbf{y})) < \eta$ and

$$-\epsilon < u(\sigma^n(\mathbf{y})) - u(\mathbf{y}) \leq -S_n(A - \beta_A)(\mathbf{y}) < \mathfrak{S}_1^{\eta, \eta}(\mathbf{x}, \mathbf{x}) + \epsilon \leq \epsilon.$$

Therefore, $\mathbf{x}$ is an Aubry point.

It remains to argue that $h_A(\mathbf{x}, \mathbf{x}) = 0$ whenever $\mathbf{x} \in \Omega(A)$. But it follows immediately from Corollary 4.5 that, for any Aubry point $\mathbf{x}$,

$$\mathfrak{S}_l^{\epsilon, \epsilon}(\mathbf{x}, \mathbf{x}) < \epsilon \qquad \qquad \forall \epsilon > 0, \ \forall l \geq 0. \tag{5.7}$$

Thus, we conclude that $h_A(\mathbf{x}, \mathbf{x}) \leq 0$ or, better yet, $h_A(\mathbf{x}, \mathbf{x}) = 0$.

Item iv. Let us start showing that $h_A(\mathbf{x}, \cdot) = \phi_A(\mathbf{x}, \cdot)$ when $\mathbf{x} \in \Omega(A)$. Combining (5.6) and (5.7), we have

$$\mathfrak{S}_l^{\lambda^k, \lambda^k}(\mathbf{x}, \mathbf{y}) < \lambda^k + \mathfrak{S}_1^{\lambda^k, \lambda^k}(\mathbf{x}, \mathbf{y}) + \frac{\mathrm{Lip}(A)}{1 - \lambda} \lambda^k, \qquad \forall \mathbf{x} \in \Omega(A), \ \forall \mathbf{y} \in \Sigma.$$

Taking the supremum over l and then passing to the limit as k tends to infinity, we get $h_A(\mathbf{x}, \mathbf{y}) \leq \phi_A(\mathbf{x}, \mathbf{y})$. Hence, $h_A(\mathbf{x}, \cdot) = \phi_A(\mathbf{x}, \cdot)$ if $\mathbf{x} \in \Omega(A)$.

Let us now argue that $\phi_A(\mathbf{x}, \cdot) = h_A(\mathbf{x}, \cdot)$ is a well-defined real-valued function if $\mathbf{x} \in \Omega(A)$. Thanks to item i, it suffices to show that $h_A(\mathbf{x}, \cdot) < \infty$ for $\mathbf{x} \in \Omega(A)$. Since (Σ, σ) is a topologically mixing dynamical system, we remark that $\mathfrak{S}_0^{\lambda, 0}(\bar{\mathbf{x}}, \bar{\mathbf{y}}) < \infty$ for all $\bar{\mathbf{x}}, \bar{\mathbf{y}} \in \Sigma$. Notice then that, from the inequality (5.2) with $k' = \bar{k} = 1$, $l \geq k - 1$, $\bar{l} = 0$ and $\bar{\epsilon}' = 0$, combined with the inequality (5.7), we obtain

$$\mathfrak{S}_l^{\lambda^k,0}(\mathbf{x},\mathbf{y}) < \lambda^k + \mathfrak{S}_0^{\lambda,0}(\mathbf{x},\mathbf{y}) + \frac{\mathrm{Lip}(A)}{1-\lambda}\lambda, \qquad \forall\,\mathbf{x}\in\Omega(A),\,\forall\,\mathbf{y}\in\Sigma.$$

Given points $\bar{\mathbf{x}},\bar{\mathbf{y}}\in\Sigma$, obviously $\mathfrak{S}_0^{\lambda^k,0}(\bar{\mathbf{x}},\bar{\mathbf{y}})\leq\cdots\leq\mathfrak{S}_{l-1}^{\lambda^k,0}(\bar{\mathbf{x}},\bar{\mathbf{y}})\leq\mathfrak{S}_l^{\lambda^k,0}(\bar{\mathbf{x}},\bar{\mathbf{y}})$ and $\mathfrak{S}_\ell^{\lambda^k,\lambda^k}(\bar{\mathbf{x}},\bar{\mathbf{y}})\leq\mathfrak{S}_\ell^{\lambda^k,0}(\bar{\mathbf{x}},\bar{\mathbf{y}})$ for $\ell\geq 0$. Thus, from the above inequality, we deduce that $h_A(\mathbf{x},\mathbf{y})\leq\mathfrak{S}_0^{\lambda,0}(\mathbf{x},\mathbf{y})+\frac{\lambda}{1-\lambda}\mathrm{Lip}(A)<\infty$ for $\mathbf{x}\in\Omega(A)$ and $\mathbf{y}\in\Sigma$.

Let us prove that the real-valued function $h_A(\mathbf{x},\cdot)=\phi_A(\mathbf{x},\cdot)$, $\mathbf{x}\in\Omega(A)$, is Lipschitz continuous. Given any points $\bar{\mathbf{x}},\bar{\mathbf{y}}\in\Sigma$ with $d(\bar{\mathbf{x}},\bar{\mathbf{y}})\leq\lambda$, we apply the inequality (5.2) with $\lambda^{\bar{k}}=d(\bar{\mathbf{x}},\bar{\mathbf{y}})$, $k=k'\geq\bar{k}$, $\bar{\ell}'=\lambda^k$, $l\geq k-1$ and $\bar{l}=0$ so that

$$\mathfrak{S}_l^{\lambda^k,\lambda^k}(\mathbf{x},\bar{\mathbf{y}})\leq\mathfrak{S}_l^{\lambda^k,\lambda^k}(\mathbf{x},\bar{\mathbf{x}})+\mathfrak{S}_0^{d(\bar{\mathbf{x}},\bar{\mathbf{y}}),\lambda^k}(\bar{\mathbf{x}},\bar{\mathbf{y}})+\frac{\mathrm{Lip}(A)}{1-\lambda}d(\bar{\mathbf{x}},\bar{\mathbf{y}}).$$

Clearly $\mathfrak{S}_0^{d(\bar{\mathbf{x}},\bar{\mathbf{y}}),\lambda^k}(\bar{\mathbf{x}},\bar{\mathbf{y}})\leq -S_0(A-\beta_A)(\bar{\mathbf{y}})=0$. Furthermore, since $\mathfrak{S}_0^{\lambda^k,\lambda^k}(\mathbf{x},\bar{\mathbf{y}})\leq\mathfrak{S}_1^{\lambda^k,\lambda^k}(\mathbf{x},\bar{\mathbf{y}})\leq\cdots\leq\mathfrak{S}_l^{\lambda^k,\lambda^k}(\mathbf{x},\bar{\mathbf{y}})$, from the above inequality we have

$$\mathfrak{S}_\ell^{\lambda^k,\lambda^k}(\mathbf{x},\bar{\mathbf{y}})\leq\sup_l\mathfrak{S}_l^{\lambda^k,\lambda^k}(\mathbf{x},\bar{\mathbf{x}})+\frac{\mathrm{Lip}(A)}{1-\lambda}d(\bar{\mathbf{x}},\bar{\mathbf{y}}),\qquad\forall\,\ell\geq 0,$$

which immediately yields

$$h_A(\mathbf{x},\bar{\mathbf{y}})-h_A(\mathbf{x},\bar{\mathbf{x}})\leq\frac{\mathrm{Lip}(A)}{1-\lambda}d(\bar{\mathbf{x}},\bar{\mathbf{y}}). \tag{5.8}$$

Since $\bar{\mathbf{x}}$ and $\bar{\mathbf{y}}$ play symmetric roles, we conclude that $h_A(\mathbf{x},\cdot)=\phi_A(\mathbf{x},\cdot)$, $\mathbf{x}\in\Omega(A)$, is Lipschitz continuous.

In order to see that $\phi_A(\mathbf{x},\cdot)=h_A(\mathbf{x},\cdot)$, $\mathbf{x}\in\Omega(A)$, is a sub-action for A, just note that the inequality (5.3) and the very definition of the Mañé potential ensure

$$\phi_A(\mathbf{x},\sigma(\bar{\mathbf{x}}))-\phi_A(\mathbf{x},\bar{\mathbf{x}})\leq\phi_A(\bar{\mathbf{x}},\sigma(\bar{\mathbf{x}}))\leq-(A-\beta_A)(\bar{\mathbf{x}}).$$

Finally, let us prove that $\phi_A(\mathbf{x},\cdot)=h_A(\mathbf{x},\cdot)$, $\mathbf{x}\in\Omega(A)$, is actually a calibrated sub-action. Since we already know that we are dealing with a sub-action, it suffices to show that

$$\phi_A(\mathbf{x},\bar{\mathbf{x}})\geq\phi_A(\mathbf{x},\bar{\mathbf{y}})-A(\bar{\mathbf{y}})+\beta_A\quad\text{for some }\bar{\mathbf{y}}\in\Sigma\text{ with }\sigma(\bar{\mathbf{y}})=\bar{\mathbf{x}}.$$

As $\mathbf{x}$ is an Aubry point, given an integer $k\geq 1$, there exist $\mathbf{z}^k\in\Sigma$ and $n_k>1$ such that $d(\mathbf{z}^k,\mathbf{x})\leq\lambda^k$, $d(\sigma^{n_k}(\mathbf{z}^k),\bar{\mathbf{x}})\leq\lambda^k$ and

$$-S_{n_k}(A-\beta_A)(\mathbf{z}^k)<\mathfrak{S}_2^{\lambda^k,\lambda^k}(\mathbf{x},\bar{\mathbf{x}})+\frac{1}{k}<\lambda^k+\mathfrak{S}_1^{\lambda^k,\lambda^k}(\mathbf{x},\bar{\mathbf{x}})+\frac{\mathrm{Lip}(A)}{1-\lambda}\lambda^k+\frac{1}{k}.$$

Let $\bar{\mathbf{y}}\in\Sigma$ be a limit of a subsequence $\{\sigma^{n_{k_j}-1}(\mathbf{z}^{k_j})\}_{j>0}$. Given $\epsilon>0$, for j large enough, we obtain

$$\mathfrak{S}_1^{\epsilon,\epsilon}(\mathbf{x},\bar{\mathbf{y}}) - \left[A\left(\sigma^{n_{k_j}-1}(\mathbf{z}^{k_j})\right) - \beta_A \right]$$

$$\leq -S_{n_{k_j}-1}(A - \beta_A)(\mathbf{z}^{k_j}) - \left[A \circ \sigma^{n_{k_j}-1}(\mathbf{z}^{k_j}) - \beta_A \right]$$

$$< \mathfrak{S}_1^{\lambda^{k_j},\lambda^{k_j}}(\mathbf{x},\bar{\mathbf{x}}) + \left(\frac{\mathrm{Lip}(A)}{1-\lambda} + 1 \right) \lambda^{k_j} + \frac{1}{k_j}.$$

Passing to the limit and then considering $\epsilon > 0$ arbitrarily small, we thus get $\phi_A(\mathbf{x},\bar{\mathbf{y}}) - A(\bar{\mathbf{y}}) + \beta_A \leq \phi_A(\mathbf{x},\bar{\mathbf{x}})$. Besides, $\sigma(\bar{\mathbf{y}}) = \lim_{j\to\infty} \sigma^{n_{k_j}}(\mathbf{z}^{k_j}) = \bar{\mathbf{x}}$. □

As we have already mentioned, the Mañé potential and the Peierls barrier are concepts inspired by similar notions in Aubry-Mather theory for Lagrangian systems (see [37, 80, 81]). For instance, given a compact manifold X, for $x, y \in X$ and $t > 0$, let $C_t(x, y)$ denote the set of all absolutely continuous curves $\gamma : [0, t] \to X$ with $\gamma(0) = x$ and $\gamma(t) = y$. Notice that a closed curve is just an element $\gamma \in C_t(x, x)$ for some $x \in X$ and a particular $t > 0$. So if $L : TX \to \mathbb{R}$ is an autonomous, strictly convex and superlinear C^2 Lagrangian, recalling that the Mañé's critical value is defined as the constant

$$c(L) := \inf \left\{ c \in \mathbb{R} : \int_0^t L(\gamma(s), \dot{\gamma}(s))\, ds + ct \geq 0, \text{ for all closed curve } \gamma \right\},$$
(5.9)

one may introduce the Peierls barrier $\mathbf{h}_L : X \times X \to \mathbb{R}$ as

$$\mathbf{h}_L(x, y) := \liminf_{t\to\infty} \inf_{\gamma \in C_t(x,y)} \left[\int_0^t L(\gamma(s), \dot{\gamma}(s))\, ds + c(L)t \right].$$

For the reader interested in details, we remember that main references on Lagrangian Aubry-Mather theory are [36, 45].

In Lagrangian Aubry-Mather theory, it is well known that, for any point $x \in X$, the application $y \in X \mapsto \mathbf{h}_L(x, y) \in \mathbb{R}$ defines a viscosity solution of the Hamilton-Jacobi equation. Since we propose to consider viscosity solutions as a notion similar to our calibrated sub-actions, we can wonder if an analogous result holds in ergodic optimization. In other terms, for a Lipschitz continuous potential A, does $h_A(\mathbf{x}, \cdot)$ define a calibrated sub-action for any point $\mathbf{x} \in \Sigma$? The answer is no, and follows from the proposition below.

Proposition 5.3 *Assume that the potential A is Lipschitz continuous. If $\mathbf{x}, \mathbf{y} \in \Sigma$ and $0 < N < \min\{n > 0 : \sigma^n(\mathbf{x}) = \mathbf{y}\} \leq \infty$, then*

$$\phi_A(\mathbf{x}, \mathbf{y}) = \phi_A(\mathbf{x}, \sigma^N(\mathbf{x})) + \phi_A(\sigma^N(\mathbf{x}), \mathbf{y}).$$

Moreover, for all $\mathbf{x} \in \Sigma$ and all $N \geq 1$,

$$\phi_A(\mathbf{x}, \sigma^N(\mathbf{x})) = -S_N(A - \beta_A)(\mathbf{x}).$$

This result is based on Proposition 3.5 of [39]. Before providing the proof of this result, let us see how it shows that the Mañé potential (and consequently the Peierls barrier) does not always define a continuous function. As a matter of fact, it is easy to exhibit simple examples satisfying

$$\infty = \lim_{N \to \infty} \phi_A(\mathbf{x}, \sigma^N(\mathbf{x})) = \lim_{N \to \infty} h_A(\mathbf{x}, \sigma^N(\mathbf{x})),$$

which shows that both functions $\phi_A(\mathbf{x}, \cdot)$ and $h_A(\mathbf{x}, \cdot)$ are not continuous, since Σ is compact. In order to be completely explicit, consider, for instance, $\Sigma = \{0, 1\}^{\mathbb{N}}$ and $A = -d(\underline{0}, \cdot)$, where $\underline{0}$ indicates the fixed point $(0, 0, 0, \ldots)$. Note that $\beta_A = 0$ and $\Omega(A) = \{\underline{0}\}$. For the point

$$\mathbf{x} = (0, 1, 0, 0, 1, 0, 0, 0, 1, 0, 0, 0, 0, 1, 0, 0, 0, 0, 0, 1, \ldots) \in \Sigma,$$

applying the previous proposition, we have $\phi_A(\mathbf{x}, \sigma^N(\mathbf{x})) = -S_N A(\mathbf{x}) \geq 0$. Since $d(\sigma^N(\mathbf{x}), \underline{0}) = 1$ for infinitely many integers $N > 0$, it clearly happens

$$\lim_{N \to \infty} \phi_A(\mathbf{x}, \sigma^N(\mathbf{x})) = \infty,$$

despite that $\liminf_{N \to \infty} d(\Omega(A), \sigma^N(\mathbf{x})) = \liminf_{N \to \infty} d(\underline{0}, \sigma^N(\mathbf{x})) = 0$.

Proof of Proposition 5.3 Consider $\mathbf{x}, \mathbf{y} \in \Sigma$ and $N > 0$ as in the statement. Fix $\epsilon > 0$ satisfying $2\epsilon < \Delta := \min\{d(\sigma^k(\mathbf{x}), \mathbf{y}) : 0 < k \leq N\}$. Let then $\eta \in (0, \epsilon)$ be a constant such that, if $\mathbf{z} \in \Sigma$ verifies $d(\mathbf{x}, \mathbf{z}) < \eta$, then $d(\sigma^k(\mathbf{x}), \sigma^k(\mathbf{z})) < \epsilon$ for all $k \in \{0, 1, \ldots, N\}$. If $\mathfrak{S}_1^{\eta, \eta}$ is the auxiliary function given by (5.1), there exist a point $\mathbf{z} \in \Sigma$ and an integer $M \geq 1$ such that $d(\mathbf{z}, \mathbf{x}) < \eta$, $d(\sigma^M(\mathbf{z}), \mathbf{x}) < \eta$ and

$$-S_M(A - \beta_A)(\mathbf{z}) < \mathfrak{S}_1^{\eta, \eta}(\mathbf{x}, \mathbf{y}) + \eta.$$

We claim that $M > N$. As a matter of fact, for any integer $0 < k \leq N$, notice that $d(\sigma^k(\mathbf{z}), \mathbf{y}) \geq d(\sigma^k(\mathbf{x}), \mathbf{y}) - d(\sigma^k(\mathbf{x}), \sigma^k(\mathbf{z})) > \Delta - \epsilon > \epsilon > \eta$. Thus, in particular, we have

$$\mathfrak{S}_1^{\epsilon, \epsilon}(\sigma^N(\mathbf{x}), \mathbf{y}) \leq -S_{M-N}(A - \beta_A)(\sigma^N(\mathbf{z})).$$

The very definition of η ensures that

$$-S_N(A - \beta_A)(\mathbf{x}) \leq -S_N(A - \beta_A)(\mathbf{z}) + N\mathrm{Lip}(A)\epsilon.$$

Therefore, we get

$$\phi_A(\mathbf{x}, \sigma^N(\mathbf{x})) + \mathfrak{S}_1^{\epsilon, \epsilon}(\sigma^N(\mathbf{x}), \mathbf{y}) \leq -S_N(A - \beta_A)(\mathbf{x}) + \mathfrak{S}_1^{\epsilon, \epsilon}(\sigma^N(\mathbf{x}), \mathbf{y})$$
$$< \mathfrak{S}_1^{\eta, \eta}(\mathbf{x}, \mathbf{y}) + \eta + N\mathrm{Lip}(A)\epsilon.$$

Passing to the limit as $\eta \to 0$ and then as $\epsilon \to 0$, we obtain

$$\phi_A(\mathbf{x}, \sigma^N(\mathbf{x})) + \phi_A(\sigma^N(\mathbf{x}), \mathbf{y}) \leq -S_N(A - \beta_A)(\mathbf{x}) + \phi_A(\sigma^N(\mathbf{x}), \mathbf{y}) \leq \phi_A(\mathbf{x}, \mathbf{y}).$$

Hence, the proposition follows from the inequality (5.3).   □

5.2 Mañé Potential and Kleene Star

Historically, Lagrangian Aubry-Mather theory has been a relevant source of methods and tools which have contributed for the development of ergodic optimization. Another route, however, could have led to similar results. Ergodic optimization could have been more influenced by the theory of linear systems on dioids, being viewed as a generalized form of the spectral theory for matrices in max algebra. The first purpose of this section is thus to show to the reader a connection with this parallel and vast field, of an eminently algebraic nature, motivated by the analysis of questions arising from a number of application areas, such as production systems, transport networks, computer communication systems, etc. Comprehensive references on this subject include [8, 43, 58, 59].

We could have chosen to illustrate a link with ergodic optimization theory by exhibiting other conceptual coincidences. Indeed, in a max algebraic terminology, the ergodic maximizing value is identified with the *maximum cycle geometric mean* or *maximal circuit mean*, while the *critical graph* plays the role of the Aubry set. The reason we choose to show the relationship between the Mañé potential and the Kleene star lies in the fact that it also addresses a natural question about the representation of this action functional in the special case of locally constant potentials. In a complete idempotent semiring, the Kleene star provides a method for solving the subclass of linear equations of fixed point type (for details, see [53]). The reader interested in further connections between both theories may consult, for instance [11, 48].

So far, we have used the designation *max algebra* to indistinctly refer to theoretical aspects of two isomorphic semirings: *the max-plus algebra* and *max-times algebra*. The link between the Mañé potential and the Kleene star is easily described in terms of the first algebraic structure. By max-plus algebra, we mean the set $\mathbb{R}_{\max,+} = \mathbb{R} \cup \{-\infty\}$ with addition and multiplication given by

$$x \oplus y = \max\{x, y\}, \quad x \otimes y = x + y, \qquad \forall\, x, y \in \mathbb{R}_{\max,+}.$$

The zero element for the addition is thus $-\infty$ and the unit element for the multiplication is 0. The reader can easily check that $(\mathbb{R}_{\max,+}, \oplus, \otimes)$ is a commutative semiring. Actually it is an idempotent semifield since it is a semiring such that $x \oplus x = x$ for all x and every nonzero element has a multiplicative inverse.

The max-times algebra, by its turn, consists of the set of nonnegative real numbers $\mathbb{R}_+$ which is equipped with the same addition as above but keeps the ordinary multiplication. Note that $x \mapsto \exp x$ defines an isomorphism between both semirings.

We extend the operations $\oplus$ and $\otimes$ to matrices and vectors in the usual linear way, that is, if $A = (A(i,j))$ and $B = (B(i,j))$ are matrices with entries from $\mathbb{R}_{\max,+}$ and $\alpha \in \mathbb{R}$, then $\alpha \otimes A = (\alpha \otimes A(i,j)) = (\alpha + A(i,j))$, and, whenever A and B have compatible sizes, $A \oplus B = (A(i,j) \oplus B(i,j)) = (\max\{A(i,j), B(i,j)\})$ as well as $A \otimes B = (\bigoplus_k A(i,k) \otimes B(k,j)) = (\max_k\{A(i,k) + B(k,j)\})$. In the context of square max-plus matrices, let Id denote the corresponding identity matrix, namely, the matrix with 0 on the diagonal and $-\infty$ elsewhere. For a square max-plus matrix A, as usual A^n represents the iterated product $A \otimes A \otimes \ldots \otimes A$ with n terms. We also set $A^0 = Id$.

Given an $M \times M$ max-plus matrix $A = (A(i,j))$, we canonically associate a directed graph whose set of vertices is the set of M indices and set of edges is formed by the pairs (i,j) such that $A(i,j) \neq -\infty$. To each edge (i,j) we then assign the weight $A(i,j)$. By the weight of a given path in this weighted graph, as usual we mean the sum of the weights of the traversed edges.

We may now introduce the Kleene star. The next result is well known in the max-plus literature (see, for instance [8]).

Proposition-Definition *Let $A \in (\mathbb{R}_{\max,+})^{M \times M}$. The infinite series*

$$\bigoplus_{n \geq 0} A^n = Id \oplus A \oplus A^2 \oplus \ldots$$

converges to a matrix with entries from $\mathbb{R}_{\max,+}$ if, and only if, there is no cycle with positive weight in the associated graph of A. In such case, the limit is

$$A^* := Id \oplus A \oplus A^2 \oplus \cdots \oplus A^{M-1},$$

which is called the Kleene star of A.

Note that the entry (i,j) of the series $\bigoplus_{n \geq 0} A^n$ can be interpreted, in terms of the associated graph, as the maximum weight of a path of any length from i to j. By the very definition of the Mañé potential, this observation should guide us in establishing the desired connection.

In order to be concrete, let $A : \Sigma \to \mathbb{R}$ be a potential that depends on $m+1$ coordinates. By writing $A(\mathbf{x}) = A(x_0, x_1, \ldots, x_m) = A((x_0, \ldots, x_{m-1}), (x_1, \ldots, x_m))$, A can be clearly identified with a matrix that belongs to $(\mathbb{R}_{\max,+})^{M \times M}$, where M is the number of **M**-allowed words of length m. Indeed, if $(x_0, \ldots, x_{m-1})$ and $(y_0, \ldots, y_{m-1})$ are two **M**-allowed words but $(x_1, \ldots, x_{m-1}) \neq (y_0, \ldots, y_{m-2})$, we just set $A((x_0, \ldots, x_{m-1}), (y_0, \ldots, y_{m-1})) = -\infty$. Note now that we can rewrite the corresponding auxiliary function (5.1) as follows

$$-\mathfrak{S}_1^{\lambda^m, \lambda^m}(\mathbf{x}, \mathbf{y}) = -\mathfrak{S}_1^{\lambda^m, \lambda^m}((x_0, \ldots, x_{m-1}), (y_0, \ldots, y_{m-1}))$$

$$= \sup_{\substack{n \geq 1 \\ (w_0,\ldots,w_{m-1})=(x_0,\ldots,x_{m-1}) \\ (w_n,\ldots,w_{n+m-1})=(y_0,\ldots,y_{m-1})}} \max \quad S_n(A - \beta_A)(\mathbf{w})$$

$$= \left(\bigoplus_{n \geq 1} \left(\beta_A^{\ominus 1} \otimes A \right)^n \right) \left((x_0,\ldots,x_{m-1}),(y_0,\ldots,y_{m-1}) \right),$$

which yields an interesting equation in $(\mathbb{R}_{\max,+})^{M \times M}$

$$Id \oplus \left(-\mathfrak{S}_1^{\lambda^m,\lambda^m} \right) = \left(\beta_A^{\ominus 1} \otimes A \right)^*.$$

Recalling that $\phi_A = \lim_{\epsilon \to 0} \mathfrak{S}_1^{\epsilon,\epsilon}$, as any potential that depends on $m + 1$ coordinates may be seen as depending on $m + l$ coordinates for all $l \geq 1$, it is quite tempting to pass to the limit on the above equation and consider the following (formal) expression in $(\mathbb{R}_{\max,+})^{\Sigma \times \Sigma}$

$$Id \oplus (-\phi_A) = \left(\beta_A^{\ominus 1} \otimes A \right)^*.$$

Since $Id : \Sigma \times \Sigma \to \mathbb{R}_{\max,+}$ is just the function with 0 on the diagonal and $-\infty$ elsewhere, the left-hand side has a precise meaning and could be used to extend the Kleene star to such a context.

Chapter 6
Representation of Calibrated Sub-actions

Our purpose now is to show that the variational tools developed in the two previous chapters (the Aubry set and the action potentials between points) can be used to describe all solutions of the Lax-Oleinik fixed point problem, that is, all calibrated sub-actions. The side effects of the analysis leading to such a result are the beginning of a deeper understanding of the nature of the sub-actions themselves.

6.1 The Mañé-Peierls Transform

We show in this chapter that any continuous calibrated sub-action for a Lipschitz continuous potential is characterized by its values on the Aubry set and the values of the Mañé potential or the Peierls barrier (see Corollary 6.3). With this, in a sense, we complete the study started in Chap. 3. As already mentioned, the representation of calibrated sub-actions to be detailed here is inspired by a similar result for weak KAM solutions due to Contreras (see [34]). The next concept will be essential in the sequel.

Definition 6.A Let $A : \Sigma \to \mathbb{R}$ be a Lipschitz continuous potential. We call Mañé-Peierls transform the application $\mathcal{F}_A$ defined on the space of bounded below functions $\psi : \Omega(A) \to \mathbb{R}$ by

$$\mathcal{F}_A(\psi)(\mathbf{y}) = \inf_{\mathbf{x} \in \Omega(A)} [\psi(\mathbf{x}) + \phi_A(\mathbf{x}, \mathbf{y})] = \inf_{\mathbf{x} \in \Omega(A)} [\psi(\mathbf{x}) + h_A(\mathbf{x}, \mathbf{y})], \quad \forall \mathbf{y} \in \Sigma.$$

By item *iii* of Proposition 5.2, $\mathcal{F}_A(\psi) \leq \psi$ everywhere on $\Omega(A)$. Moreover, since ψ is bounded below, thanks to the item *i* of Proposition 5.2, we conclude that $\mathcal{F}_A(\psi)$ is a real-valued function defined on Σ. Concerning its regularity, we have the following result.

© The Author(s) 2017
E. Garibaldi, *Ergodic Optimization in the Expanding Case*,
SpringerBriefs in Mathematics, DOI 10.1007/978-3-319-66643-3_6

Lemma 6.1 *Assume that $A : \Sigma \to \mathbb{R}$ is a Lipschitz continuous potential. For any bounded below function $\psi : \Omega(A) \to \mathbb{R}$, its Mañé-Peierls transform $\mathcal{F}_A(\psi) : \Sigma \to \mathbb{R}$ is Lipschitz continuous.*

Proof Given $\epsilon > 0$ and $\bar{\mathbf{x}} \in \Sigma$, let $\mathbf{x} \in \Omega(A)$ be such that $\psi(\mathbf{x}) + h_A(\mathbf{x}, \bar{\mathbf{x}}) < \mathcal{F}_A(\psi)(\bar{\mathbf{x}}) + \epsilon$. Let $\bar{\mathbf{y}} \in \Sigma$ be any point satisfying $d(\bar{\mathbf{x}}, \bar{\mathbf{y}}) \leq \lambda$. Therefore, from item *iv* of Proposition 5.2 and particularly from (5.8), we get

$$\mathcal{F}_A(\psi)(\bar{\mathbf{y}}) - \mathcal{F}_A(\psi)(\bar{\mathbf{x}}) < h_A(\mathbf{x}, \bar{\mathbf{y}}) - h_A(\mathbf{x}, \bar{\mathbf{x}}) + \epsilon \leq \frac{\mathrm{Lip}(A)}{1 - \lambda} d(\bar{\mathbf{x}}, \bar{\mathbf{y}}) + \epsilon.$$

Since $\epsilon > 0$ is arbitrary, the previous inequality ensures that $\mathcal{F}_A(\psi)$ is a Lipschitz continuous function. $\qquad\qquad\qquad\qquad\qquad\qquad\qquad\qquad\qquad\qquad\qquad\qquad\qquad\square$

The next theorem gathers central properties of the Mañé-Peierls transform.

Theorem 6.2 *Let $A : \Sigma \to \mathbb{R}$ be a Lipschitz continuous potential. Suppose that $\psi : \Omega(A) \to \mathbb{R}$ is continuous. Then*

  i. *$\mathcal{F}_A(\psi)$ is a continuous calibrated sub-action for A;*
 ii. *if $\psi(\mathbf{y}) - \psi(\mathbf{x}) \leq \phi_A(\mathbf{x}, \mathbf{y})$ for all $\mathbf{x}, \mathbf{y} \in \Omega(A)$, then $\mathcal{F}_A(\psi)|_{\Omega(A)} = \psi$;*
iii. *if $u \in C(\Sigma)$ is any sub-action for A, then $u \leq \mathcal{F}_A(u|_{\Omega(A)})$ everywhere on Σ; moreover, if u is calibrated, then $u = \mathcal{F}_A(u|_{\Omega(A)})$;*
 iv. *the Mañé-Peierls transform $\mathcal{F}_A$ is a bijective and isometric correspondence between the functions $\psi \in C(\Omega(A))$ verifying $\psi(\mathbf{y}) - \psi(\mathbf{x}) \leq \phi_A(\mathbf{x}, \mathbf{y})$ for all $\mathbf{x}, \mathbf{y} \in \Omega(A)$ and the continuous calibrated sub-actions for A.*

Proof
Item i. Let us see that $\mathcal{F}_A(\psi) \in C(\Sigma)$ is a sub-action for the potential A. Given $\epsilon > 0$ and $\bar{\mathbf{x}} \in \Sigma$, there exists $\mathbf{x} \in \Omega(A)$ such that $\psi(\mathbf{x}) + \phi_A(\mathbf{x}, \bar{\mathbf{x}}) < \mathcal{F}_A(\psi)(\bar{\mathbf{x}}) + \epsilon$. Hence, by the item *iv* of Proposition 5.2, we have

$$\mathcal{F}_A(\psi)(\sigma(\bar{\mathbf{x}})) - \mathcal{F}_A(\psi)(\bar{\mathbf{x}}) < \phi_A(\mathbf{x}, \sigma(\bar{\mathbf{x}})) - \phi_A(\mathbf{x}, \bar{\mathbf{x}}) + \epsilon \leq \beta_A - A(\bar{\mathbf{x}}) + \epsilon.$$

By taking ϵ arbitrarily small, we obtain $A + \mathcal{F}_A(\psi) \circ \sigma - \mathcal{F}_A(\psi) \leq \beta_A$ everywhere on Σ. Actually, $\mathcal{F}_A(\psi) \in C(\Sigma)$ is a calibrated sub-action. Indeed, we only have to argue that, for any $\bar{\mathbf{x}} \in \Sigma$, we can find $\bar{\mathbf{y}} \in \Sigma$ with $\sigma(\bar{\mathbf{y}}) = \bar{\mathbf{x}}$ such that

$$\mathcal{F}_A(\psi)(\bar{\mathbf{x}}) \geq \mathcal{F}_A(\psi)(\bar{\mathbf{y}}) - A(\bar{\mathbf{y}}) + \beta_A$$

Hence, for each $k > 0$, let $\mathbf{x}^k$ be an Aubry point that satisfies $\psi(\mathbf{x}^k) + \phi_A(\mathbf{x}^k, \bar{\mathbf{x}}) < \mathcal{F}_A(\psi)(\bar{\mathbf{x}}) + \frac{1}{k}$. Since $\phi_A(\mathbf{x}^k, \cdot)$ defines a calibrated sub-action, there exists $\bar{\mathbf{y}}^k \in \Sigma$ with $\sigma(\bar{\mathbf{y}}^k) = \bar{\mathbf{x}}$ such that $\phi_A(\mathbf{x}^k, \bar{\mathbf{x}}) = \phi_A(\mathbf{x}^k, \bar{\mathbf{y}}^k) - A(\bar{\mathbf{y}}^k) + \beta_A$. Let $\bar{\mathbf{y}} \in \Sigma$ be an accumulation point of the sequence $\{\bar{\mathbf{y}}^k\}$. Obviously, $\sigma(\bar{\mathbf{y}}) = \bar{\mathbf{x}}$. Moreover,

$$\mathcal{F}_A(\psi)(\bar{\mathbf{y}}^k) - A(\bar{\mathbf{y}}^k) + \beta_A \leq \psi(\mathbf{x}^k) + \phi_A(\mathbf{x}^k, \bar{\mathbf{y}}^k) - A(\bar{\mathbf{y}}^k) + \beta_A$$

$$= \psi(\mathbf{x}^k) + \phi_A(\mathbf{x}^k, \bar{\mathbf{x}}) < \mathcal{F}_A(\psi)(\bar{\mathbf{x}}) + \frac{1}{k}, \qquad \forall\, k > 0.$$

Therefore, $\mathcal{F}_A(\psi)(\bar{\mathbf{y}}) - A(\bar{\mathbf{y}}) + \beta_A \leq \mathcal{F}_A(\psi)(\bar{\mathbf{x}})$.

Item ii. Suppose $\psi(\mathbf{y}) - \psi(\mathbf{x}) \leq \phi_A(\mathbf{x}, \mathbf{y})$ for all $\mathbf{x}, \mathbf{y} \in \Omega(A)$. Then we clearly have the inequalities

$$\mathcal{F}_A(\psi)(\mathbf{y}) \leq \psi(\mathbf{y}) \leq \psi(\mathbf{x}) + \phi_A(\mathbf{x}, \mathbf{y}), \quad \forall\, \mathbf{x}, \mathbf{y} \in \Omega(A).$$

By taking the infimum over $\mathbf{x} \in \Omega(A)$, one concludes that $\mathcal{F}_A(\psi) = \psi$ everywhere on $\Omega(A)$.

Item iii. Thanks to item i of Proposition 5.2, given $\mathbf{y} \in \Sigma$, we verify

$$u(\mathbf{y}) \leq \inf_{\mathbf{x} \in \Omega(A)} [u(\mathbf{x}) + \phi_A(\mathbf{x}, \mathbf{y})] = \mathcal{F}_A(u|_{\Omega(A)})(\mathbf{y}).$$

If $\{\bar{\mathbf{x}}^k\} \subset \Sigma$ is a u-optimal trajectory with $\bar{\mathbf{x}}^0 = \mathbf{y}$, let $\bar{\mathbf{x}} \in \Sigma$ be an accumulation point of $\{\bar{\mathbf{x}}^k\}$. According to the Proposition 4.1, $\bar{\mathbf{x}}$ is an Aubry point. We claim that $u(\mathbf{y}) = u(\bar{\mathbf{x}}) + \phi_A(\bar{\mathbf{x}}, \mathbf{y})$. Indeed,

$$u(\mathbf{y}) = u(\bar{\mathbf{x}}^k) - \sum_{j=0}^{k} (A - \beta_A)(\bar{\mathbf{x}}^j) = u(\bar{\mathbf{x}}^k) - S_k(A - \beta_A)(\bar{\mathbf{x}}^k).$$

Given $\epsilon > 0$, there exists $\eta \in (0, \epsilon)$ such that, for $\mathbf{x}', \mathbf{y}' \in \Sigma$, $d(\mathbf{x}', \mathbf{y}') < \eta$ implies $|u(\mathbf{x}') - u(\mathbf{y}')| < \epsilon$. If $\mathfrak{S}_1^{\epsilon,\epsilon}$ is the auxiliary function given by (5.1), for $m > 0$ large enough so that $d(\bar{\mathbf{x}}, \bar{\mathbf{x}}^{k_m}) < \eta$, we get

$$u(\mathbf{y}) = u(\bar{\mathbf{x}}^{k_m}) - S_{k_m}(A - \beta_A)(\bar{\mathbf{x}}^{k_m}) > u(\bar{\mathbf{x}}) - \epsilon + \mathfrak{S}_1^{\epsilon,\epsilon}(\bar{\mathbf{x}}, \mathbf{y}).$$

Since $\epsilon > 0$ is arbitrary, we have $u(\mathbf{y}) \geq u(\bar{\mathbf{x}}) + \phi_A(\bar{\mathbf{x}}, \mathbf{y})$ with $\bar{\mathbf{x}} \in \Omega(A)$.

Item iv. The first item shows that $\mathcal{F}_A$ takes its values on the set of continuous calibrated sub-actions. It follows from item ii that $\mathcal{F}_A$ is one-to-one. Furthermore, item iii guarantees that $\mathcal{F}_A$ is onto.

Actually, the Mañé-Peierls transform is an isometry. In fact, given a point $\mathbf{y} \in \Sigma$ and a constant $\epsilon > 0$, there exists an Aubry point $\mathbf{x}$ such that $\psi(\mathbf{x}) + \phi_A(\mathbf{x}, \mathbf{y}) < \mathcal{F}_A(\psi)(\mathbf{y}) + \epsilon$. Thus,

$$\mathcal{F}_A(\hat{\psi})(\mathbf{y}) - \mathcal{F}_A(\psi)(\mathbf{y}) \leq \hat{\psi}(\mathbf{x}) - \psi(\mathbf{x}) + \epsilon \leq \|\hat{\psi} - \psi\|_\infty + \epsilon.$$

As ϵ tends to 0, since $\mathbf{y}$ is arbitrary and ψ and $\hat{\psi}$ play symmetrical roles, we obtain $\|\mathcal{F}_A(\psi) - \mathcal{F}_A(\hat{\psi})\|_\infty \leq \|\psi - \hat{\psi}\|_\infty$. On the other hand, obviously $\mathcal{F}_A(\psi)|_{\Omega(A)} = \psi$ and $\mathcal{F}_A(\hat{\psi})|_{\Omega(A)} = \hat{\psi}$ imply the opposite inequality $\|\mathcal{F}_A(\psi) - \mathcal{F}_A(\hat{\psi})\|_\infty \geq \|\psi - \hat{\psi}\|_\infty$. $\qquad\square$

We may point out several straightforward corollaries.

Corollary 6.3 *Let $u \in C(\Sigma)$ be a calibrated sub-action for a Lipschitz continuous potential $A : \Sigma \to \mathbb{R}$. Then the following representation formula holds*

$$u(\mathbf{y}) = \min_{\mathbf{x} \in \Omega(A)} [u(\mathbf{x}) + \phi_A(\mathbf{x}, \mathbf{y})] = \min_{\mathbf{x} \in \Omega(A)} [u(\mathbf{x}) + h_A(\mathbf{x}, \mathbf{y})], \quad \forall\, \mathbf{y} \in \Sigma.$$

Corollary 6.4 *Let $u, v \in C(\Sigma)$ be sub-actions for a Lipschitz continuous potential $A : \Sigma \to \mathbb{R}$. If v is calibrated and $u \leq v$ on $\Omega(A)$, then $u \leq v$ everywhere on Σ. In particular, if u and v are calibrated and $u|_{\Omega(A)} = v|_{\Omega(A)}$, then both sub-actions coincide on Σ.*

Corollary 6.5 *Let $u \in C(\Sigma)$ be an arbitrary sub-action for a Lipschitz continuous potential $A : \Sigma \to \mathbb{R}$. Then $u|_{\Omega(A)}$ is necessarily Lipschitz continuous.*

Corollary 6.6 *Let $u \in C(\Sigma)$ be an arbitrary sub-action for a Lipschitz continuous potential $A : \Sigma \to \mathbb{R}$. Then, u behaves as a calibrated sub-action on $\Omega(A)$, that is,*

$$u(\mathbf{x}) = \min_{\sigma(\mathbf{y}) = \mathbf{x}} [u(\mathbf{y}) - A(\mathbf{y}) + \beta_A], \quad \forall\, \mathbf{x} \in \Omega(A).$$

The following proposition is a well-known result in ergodic optimization (see, for instance [19]).

Proposition 6.7 *Let A be a Lipschitz continuous potential. If the restriction of σ to $\Omega(A)$ is transitive, then up to constants there is a unique continuous calibrated sub-action for A. In particular, if there exists a unique A-maximizing probability, then continuous calibrated sub-actions for A give rise to a singleton of $C(\Sigma)/\mathbb{R}$.*

Proof Let $\bar{\mathbf{x}}$ be a point of $\Omega(A)$ whose orbit is dense. For $u, v \in C(\Sigma)$ calibrated sub-actions, thanks to (4.2), we have

$$u(\bar{\mathbf{x}}) - u(\sigma^n(\bar{\mathbf{x}})) = S_n(A - \beta_A)(\bar{\mathbf{x}}) = v(\bar{\mathbf{x}}) - v(\sigma^n(\bar{\mathbf{x}})), \quad \forall\, n \geq 1.$$

Hence, given $\mathbf{x}$ an Aubry point, by considering a suitable subsequence of $\{\sigma^n(\bar{\mathbf{x}})\}$, we obtain at the limit $u(\bar{\mathbf{x}}) - u(\mathbf{x}) = v(\bar{\mathbf{x}}) - v(\mathbf{x})$, which shows that $u - v$ is identically constant on $\Omega(A)$. From Corollary 6.3, this property extends to the space Σ.

Finally, as already remarked, any invariant probability whose support lies on $\Omega(A)$ is a maximizing measure. Therefore, if the Aubry set had two transitive pieces, there would be at least two maximizing probabilities. □

The above proposition can be applied, for instance, to the Lipschitz continuous potential $A(\mathbf{x}) = -d(X, \mathbf{x})$ whenever $X \subset \Sigma$ is minimal, since, in such a case, it is easy to see that $\Omega(A) = X$. Nevertheless, in the next chapter, Proposition 7.5 will ensure that, for a dense subset of Lipschitz continuous potentials, the corresponding Aubry set coincides with the support of the unique maximizing measure, so that, at least in a topological sense, this is the typical example of application of the previous proposition. Corollary 6.3 is also useful for the analysis of the set of continuous calibrated sub-actions in richer situations (see [51]).

Chapter 7
Separating Sub-actions

We focus in this chapter on a special category of sub-actions, those for which the defining cohomological inequality becomes an equality on the smallest possible subset of the phase space, that is, on the Aubry set. Named separating sub-actions, we will show how they can be obtained from non-trivial convex combinations of the members of the family of calibrated sub-actions given by the Peierls barrier or by the Mañé potential.

7.1 A Minimalist Sub-action

A separating sub-action is other useful notion in ergodic optimization that has a counterpart in weak KAM theory for Lagrangian dynamics. In fact, a global critical subsolution of the Hamilton-Jacobi equation is the analogous concept. The existence of C^1 critical subsolutions defined on a C^∞ second countable manifold without boundary is a result due to Fathi and Siconolfi (see [46]). For compact manifolds, Bernard showed that there always exist $C^{1,1}$ global critical subsolutions (see [12]). In ergodic optimization, we focus here on the following subclass of sub-actions.

Definition 7.A Let $u \in C(\Sigma)$ be a sub-action for a continuous potential A. We say that u is separating when

$$\Omega(A) = (A + u \circ \sigma - u)^{-1}(\beta_A).$$

For Lipschitz continuous potentials, the existence of separating sub-actions and the fact that they are generic among Lipschitz continuous sub-actions was established in [51]. We provide here a new proof of their existence based on convex combinations of the calibrated sub-actions $\phi_A(\mathbf{x}, \cdot) = h_A(\mathbf{x}, \cdot)$, with $\mathbf{x} \in \Omega(A)$. It is worth noting that the Aubry set, as a separable space, admits probabilities that

© The Author(s) 2017
E. Garibaldi, *Ergodic Optimization in the Expanding Case*,
SpringerBriefs in Mathematics, DOI 10.1007/978-3-319-66643-3_7

give mass to all its open subsets. This is the case, for instance, of weighted sums of Dirac delta measures supported on points of a dense and (at most) countable subset of $\Omega(A)$.

Theorem 7.1 *Let $A : \Sigma \to \mathbb{R}$ be a Lipschitz continuous potential. For any Borel probability measure ρ on $\Omega(A)$ that gives mass to all induced open balls, the function $u_\rho : \Sigma \to \mathbb{R}$ defined by*

$$u_\rho(\mathbf{y}) := \int_{\Omega(A)} \phi_A(\mathbf{x}, \mathbf{y}) \, d\rho(\mathbf{x}) = \int_{\Omega(A)} h_A(\mathbf{x}, \mathbf{y}) \, d\rho(\mathbf{x}), \qquad \forall \, \mathbf{y} \in \Sigma, \qquad (7.1)$$

is a Lipschitz continuous separating sub-action.

This theorem was suggested by the so-called strict visualization scaling problem in max-times algebra (for details, see [29]). As we will see in the next chapter, the converse does not hold, that is, it is not possible to characterize all separating sub-actions by a representation as (7.1). The proof of Theorem 7.1 will need some auxiliary results.

Proposition 7.2 *If A is a Lipschitz continuous potential, then the following statements are equivalent:*

  i. $\mathbf{y} \in \Omega(A)$;
 ii. $\phi_A(\mathbf{x}, \sigma(\mathbf{y})) = \phi_A(\mathbf{x}, \mathbf{y}) + \phi_A(\mathbf{y}, \sigma(\mathbf{y})) \qquad \forall \, \mathbf{x} \in \Omega(A)$;
iii. $h_A(\mathbf{x}, \sigma(\mathbf{y})) = h_A(\mathbf{x}, \mathbf{y}) + h_A(\mathbf{y}, \sigma(\mathbf{y})) \qquad \forall \, \mathbf{x} \in \Omega(A)$.

Proof It is easy to see that item *iii* implies item *ii*. Moreover, since Proposition 5.3 ensures $\phi_A(\mathbf{y}, \sigma(\mathbf{y})) = -[A(\mathbf{y}) - \beta_A]$, the fact that item *i* implies item *ii* (and therefore item *iii*) comes from the inclusion (4.2) applied to the sub-action $u = \phi_A(\mathbf{x}, \cdot)$, $\mathbf{x} \in \Omega(A)$.

Suppose now that the equality of item *ii* holds for all $\mathbf{x} \in \Omega(A)$. We will show that $\mathbf{y} \in \Omega(A)$. We consider two situations.

Case 1. Either $\sigma(\mathbf{y})$ is not a periodic point.

From Proposition 5.3, for all $N > 1$, we thus obtain

$$\phi_A(\mathbf{y}, \sigma(\mathbf{y})) = \phi_A(\mathbf{y}, \sigma^N(\mathbf{y})) + \phi_A(\sigma^N(\mathbf{y}), \sigma(\mathbf{y})).$$

Given $\mathbf{x} \in \Omega(A)$, using the hypothesis, notice that

$$\phi_A(\mathbf{x}, \sigma^N(\mathbf{y})) \leq \phi_A(\mathbf{x}, \mathbf{y}) + \phi_A(\mathbf{y}, \sigma^N(\mathbf{y}))$$

$$= \phi_A(\mathbf{x}, \sigma(\mathbf{y})) - \phi_A(\sigma^N(\mathbf{y}), \sigma(\mathbf{y})) \leq \phi_A(\mathbf{x}, \sigma^N(\mathbf{y})).$$

We have shown that, for all $N \geq 1$,

$$\phi_A(\mathbf{x}, \sigma^N(\mathbf{y})) = \phi_A(\mathbf{x}, \mathbf{y}) + \phi_A(\mathbf{y}, \sigma^N(\mathbf{y})), \qquad \forall \, \mathbf{x} \in \Omega(A). \qquad (7.2)$$

Case 2. Or $\sigma(\mathbf{y})$ is a periodic point of period $M \geq 1$.

We may still apply Proposition 5.3 for $1 < N \le M$ and, in the same way, we ensure that the equation in (7.2) holds for these indexes. But then periodicity itself extends the equality for all indexes $N \ge 1$.

We claim now that, if $\mathbf{y} \in \Sigma$ is a point for which (7.2) holds for all $N \ge 1$, then $\mathbf{y} \in \Omega(A)$. Indeed, if $\mathbf{y}$ is periodic, we immediately obtain $\phi_A(\mathbf{y}, \mathbf{y}) = 0$, that is, $\mathbf{y}$ is an Aubry point. Let us thus suppose that $\mathbf{y}$ is not periodic. Recall that $\phi_A(\mathbf{y}, \sigma^N(\mathbf{y})) = -S_N(A - \beta_A)(\mathbf{y})$. Hence, since the following equalities hold for all $M \ge N$

$$
\begin{aligned}
|S_{M-N}(A - \beta_A)(\sigma^N(\mathbf{y}))| &= |\phi_A(\mathbf{y}, \sigma^N(\mathbf{y})) - \phi_A(\mathbf{y}, \sigma^M(\mathbf{y}))| \\
&= |\phi_A(\mathbf{x}, \sigma^N(\mathbf{y})) - \phi_A(\mathbf{x}, \sigma^M(\mathbf{y}))|,
\end{aligned}
$$

we see that all accumulation point of $\{\sigma^N(\mathbf{y})\}$ is an Aubry point. Let $\mathbf{z}$ be one of these accumulations points, namely, $\lim_{k \to \infty} \sigma^{N_k}(\mathbf{y}) = \mathbf{z} \in \Omega(A)$. From Proposition 5.3, we have $\phi_A(\mathbf{z}, \mathbf{y}) = \phi_A(\mathbf{z}, \sigma^{N_k}(\mathbf{y})) + \phi_A(\sigma^{N_k}(\mathbf{y}), \mathbf{y})$, which in particular shows that $\lim_{k \to \infty} \phi_A(\sigma^{N_k}(\mathbf{y}), \mathbf{y}) = \phi_A(\mathbf{z}, \mathbf{y})$. Since ϕ_A is lower semi-continuous, we get

$$
\begin{aligned}
0 \le \phi_A(\mathbf{y}, \mathbf{y}) &\le \liminf_{k \to \infty} \phi_A(\mathbf{y}, \sigma^{N_k}(\mathbf{y})) + \phi_A(\mathbf{z}, \mathbf{y}) \\
&= \lim_{k \to \infty} \phi_A(\mathbf{z}, \sigma^{N_k}(\mathbf{y})) = \phi_A(\mathbf{z}, \mathbf{z}) = 0.
\end{aligned}
$$

Therefore, $\mathbf{y} \in \Omega(A)$. $\square$

Lemma 7.3 *Let $A : \Sigma \to \mathbb{R}$ be a Lipschitz continuous potential. For any fixed point $\mathbf{y} \in \Sigma$, the map $\mathbf{x} \in \Omega(A) \mapsto \phi_A(\mathbf{x}, \mathbf{y}) = h_A(\mathbf{x}, \mathbf{y}) \in \mathbb{R}$ is Lipschitz continuous.*

Proof Given $\mathbf{x}, \bar{\mathbf{x}} \in \Omega(A)$, notice that

$$
\phi_A(\mathbf{x}, \bar{\mathbf{x}}) \ge \phi_A(\mathbf{x}, \mathbf{y}) - \phi_A(\bar{\mathbf{x}}, \mathbf{y}) \ge -\phi_A(\bar{\mathbf{x}}, \mathbf{x}).
$$

Since $\phi_A(\mathbf{x}, \mathbf{x}) = 0 = \phi_A(\bar{\mathbf{x}}, \bar{\mathbf{x}})$, the lemma follows from item *iv* of Proposition 5.2 and from Proposition 3.3. $\square$

We may now ensure that there always exist separating sub-actions for Lipschitz continuous potentials.

Proof of Theorem 7.1 Since $\phi_A(\mathbf{x}, \cdot) = h_A(\mathbf{x}, \cdot)$, with $\mathbf{x} \in \Omega(A)$, defines a family of (calibrated) sub-actions and ρ is a probability measure, it is easy to see that u_ρ is a sub-action. Moreover, thanks to Proposition 3.3, u_ρ is Lipschitz continuous.

The equality $(A + u_\rho \circ \sigma - u_\rho)(\mathbf{y}) = \beta_A$ is equivalent to

$$
A(\mathbf{y}) - \beta_A + \phi_A(\mathbf{x}, \sigma(\mathbf{y})) = \phi_A(\mathbf{x}, \mathbf{y}) \qquad \rho\text{-a.e. } \mathbf{x} \in \Omega(A).
$$

Recall that $\phi_A(\mathbf{y}, \sigma(\mathbf{y})) = -[A(\mathbf{y}) - \beta_A]$. Hence, since ρ is a measure on $\Omega(A)$ that gives mass to all induced open balls, Lemma 7.3 guarantees that

$$
\phi_A(\mathbf{x}, \sigma(\mathbf{y})) = \phi_A(\mathbf{x}, \mathbf{y}) + \phi_A(\mathbf{y}, \sigma(\mathbf{y})) \qquad \forall \mathbf{x} \in \Omega(A).
$$

From Proposition 7.2, we conclude that $\mathbf{y} \in \Omega(A)$ and therefore that u_ρ is separating. $\square$

As we have previously remarked, the existence of a separating sub-action for a potential A shows in particular that the set equality (1.5) does hold. Moreover, from (1.4) we immediately obtain the next corollary of Theorem 7.1.

Corollary 7.4 *For a Lipschitz continuous potential, an invariant probability is maximizing if, and only if, its support lies on the corresponding Aubry set.*

Another consequence is below highlighted.

Proposition 7.5 *The subset of Lipschitz continuous potentials for which the Aubry set coincides with the support of the unique (ergodic) maximizing probability is dense in the Lipschitz topology.*

It is well known that continuous potentials with a unique maximizing probability are generic in the uniform topology (for a proof, see, for instance, [40]). For the expanding case (or more generally, for the hyperbolic case), an interesting question in ergodic optimization is whether Lipschitz continuous potentials with a unique maximizing probability supported on a periodic orbit are generic in the Lipschitz topology. Partial positive answers were provided, for example, in [20, 40, 60, 85, 94]. Similar question with respect to generic dynamics was addressed in distinct contexts [1–3]. Recently, a probabilistic version of the problem was brought to the attention of the academic community [18]. A proof for the original conjecture was proposed by Contreras in [35].

Proof Suppose that $A : \Sigma \to \mathbb{R}$ is Lipschitz continuous. Thanks to the ergodic decomposition theorem, A admits an ergodic maximizing probability $\mu_A \in \mathcal{M}_\sigma$. We may assume that σ is uniquely ergodic on $\mathrm{supp}(\mu_A)$. Thus, given $\epsilon > 0$, define

$$\tilde{A} = A - \epsilon d(\mathrm{supp}(\mu_A), \cdot).$$

Obviously, $\tilde{A}$ is a Lipschitz perturbation of the potential A.

Moreover, remember that $\Omega(\tilde{A}) = \Omega(\tilde{A} + f \circ \sigma - f - c)$ for any function $f \in C(\Sigma)$ and for every constant $c \in \mathbb{R}$. So let $u : \Sigma \to \mathbb{R}$ be a Lipschitz continuous separating sub-action for A. Therefore, we have

$$\Omega(\tilde{A}) = \Omega\left(B - \epsilon d(\mathrm{supp}(\mu_A), \cdot)\right),$$

where the associated normalized potential $B := A + u \circ \sigma - u - \beta_A$ verifies both $B \le 0 = \beta_B$ and $B^{-1}(0) = \Omega(B) = \Omega(A)$. Clearly

$$\tilde{B} := B - \epsilon d(\mathrm{supp}(\mu_A), \cdot) \le 0,$$

$$\tilde{B}(\mathbf{x}) = 0 \iff B(\mathbf{x}) = 0 \text{ and } d(\mathrm{supp}(\mu_A), \mathbf{x}) = 0,$$

$$\text{and } \int \tilde{B} \, d\mu_A = 0 \implies \beta_{\tilde{B}} = 0.$$

In particular, μ_A is a $\tilde{B}$-maximizing probability and 0 is a sub-action for $\tilde{B}$. Therefore, from Corollary 7.4 and inclusion (4.2), we get

$$\text{supp}(\mu_A) \subset \Omega(\tilde{B}) \subset \tilde{B}^{-1}(0) = \text{supp}(\mu_A),$$

which ensures $\Omega(\tilde{A}) = \Omega(\tilde{B}) = \text{supp}(\mu_A)$. □

Chapter 8
Further Properties of Sub-actions

We enrich the discussion by taking into account less explored aspects of the set of continuous sub-actions, like the fact that, when considered up to constants, they form, in general, a non-compact subset of the quotient space. Such a property allows us to argue that, for Lipschitz continuous potentials that are not cohomologous to a constant, the separating sub-actions explicitly constructed in the previous chapter are quite particular and actually represent a small part of the whole set of Lipschitz continuous separating sub-actions.

8.1 Convexity, Non-compactness, and Extremal Elements

Given a continuous potential $A : \Sigma \to \mathbb{R}$, the set of continuous (respectively bounded measurable) sub-actions for A is clearly convex. Moreover, if u and v are continuous (respectively bounded measurable) sub-actions for A and $t \in (0, 1)$, then it is easy to see that the sub-action $w := tu + (1 - t)v$ satisfies

$$(A + w \circ \sigma - w)^{-1}(\beta_A) = (A + u \circ \sigma - u)^{-1}(\beta_A) \cap (A + v \circ \sigma - v)^{-1}(\beta_A).$$

In particular, a non-trivial convex combination of a separating sub-action with an arbitrary sub-action always results in a separating sub-action, which shows that, at least for Lipschitz continuous potentials, (Lipschitz) continuous separating sub-actions are dense among the (Lipschitz) continuous sub-actions.

It is plain to check that the set of continuous sub-actions is a closed subset of $(C(\Sigma), \| \cdot \|_\infty)$. We will show that the set of continuous sub-actions is non-compact for a Lipschitz continuous potential A that is not cohomologous to a constant. Actually, since the addition of a real constant does not change the role played by a sub-action, it is more convenient to discuss the question of non-compactness in the

© The Author(s) 2017
E. Garibaldi, *Ergodic Optimization in the Expanding Case*,
SpringerBriefs in Mathematics, DOI 10.1007/978-3-319-66643-3_8

quotient space $(C(\Sigma)/\mathbb{R}, \|\cdot\|_{\#})$, where recall that $\|[f]\|_{\#} = \min_{c\in\mathbb{R}} \|f+c\|_{\infty}$ for all $[f] \in C(\Sigma)/\mathbb{R}$. As we have noticed before, Proposition 3.3 ensures that continuous calibrated sub-actions form a compact subset of $(C(\Sigma)/\mathbb{R}, \|\cdot\|_{\#})$. For arbitrary sub-actions, we have the following result.

Proposition 8.1 *Let $A : \Sigma \to \mathbb{R}$ be a Lipschitz continuous potential. The following statements are equivalent:*

i. A is not cohomologous to a constant;
ii. the set of continuous sub-actions for A defines a non-compact subset of the quotient space $(C(\Sigma)/\mathbb{R}, \|\cdot\|_{\#})$.

Proof Suppose that A is cohomologous to a constant, namely, that there exists a continuous function $f : \Sigma \to \mathbb{R}$ such that $A + f \circ \sigma - f = \beta_A$. In particular, $m_A = \mathcal{M}_{\sigma}$ and $\Omega(A) = \Sigma$. Thus, it follows from Corollary 6.6 and Proposition 6.7 that, in this case, continuous sub-actions for A give rise to a singleton of $C(\Sigma)/\mathbb{R}$.

Suppose now that A is not cohomologous to a constant. From a classical result of Livsic [73], there are two periodic probabilities such that the integrals of A with respect to both measures are distinct. In particular, $\Omega(A) \cap \Omega(-A) = \emptyset$. This fact will be useful to show the non-compactness of the quotient set of continuous sub-actions.

Let $u : \Sigma \to \mathbb{R}$ be an arbitrary continuous sub-action for the potential A. Consider the associated normalized potential $B := A + u \circ \sigma - u - \beta_A \leq 0$ and finitely many continuous functions $\theta_j : \Sigma \to \mathbb{R}, j \in \{0, 1, \ldots, k\}$, satisfying $0 \leq \theta_k \leq \ldots \leq \theta_1 \leq \theta_0 \leq 1$. Define a continuous function $v_k : \Sigma \to \mathbb{R}$ by

$$v_k := u + \sum_{j=0}^{k} (\theta_j B) \circ \sigma^j.$$

Note that $v_k \leq u$ everywhere and $v_k = u$ on $\Omega(A)$. We claim v_k is a (continuous) sub-action for A. Indeed,

$$v_k \circ \sigma - v_k = u \circ \sigma - u + \sum_{j=0}^{k} (\theta_j B) \circ \sigma^{j+1} - \sum_{j=0}^{k} (\theta_j B) \circ \sigma^j$$

$$= u \circ \sigma - u + (\theta_k B) \circ \sigma^{k+1} - \theta_0 B + \sum_{j=1}^{k} [(\theta_{j-1} - \theta_j)B] \circ \sigma^j$$

$$\leq u \circ \sigma - u - B = \beta_A - A.$$

Therefore, given any continuous sub-action u for A, one may introduce a family of continuous sub-actions $\{v_k\}$ indexed by $k \geq 0$ and $\theta_j \in C(\Sigma), j \in \{0, 1, \ldots, k\}$, with $0 \leq \theta_k \leq \ldots \leq \theta_1 \leq \theta_0 \leq 1$.

We will suppose henceforth that $u \in C(\Sigma)$ is a separating sub-action for A. Thus, $\mathbf{x} \in \Omega(A)$ is equivalent to $B(\mathbf{x}) = 0$. Furthermore, in the construction of the previous family of sub-actions, we will impose a restriction: $\theta_k \geq \eta > 0$ for some positive constant η. Notice then that $v_k = u$ on $\Omega(A)$ as before, but now $v_k < u$ everywhere on $\Sigma - \Omega(A)$. In such a situation, we claim that

$$\lim_{k \to \infty} \|[v_k] - [u]\|_\# = \infty.$$

In particular, since the set of equivalence classes of continuous sub-actions will not have a finite diameter as a subset of $(C(\Sigma)/\mathbb{R}, \|\cdot\|_\#)$, its non-compactness will immediately follow. So let us determine such an infinite limit. First, note that

$$\|[v_k] - [u]\|_\# = \min_{c \in \mathbb{R}} \max_{\mathbf{x} \in \Sigma} |v_k(\mathbf{x}) - u(\mathbf{x}) + c|$$

$$\geq \min_{c \in \mathbb{R}} \max_{\mathbf{x} \in \Omega(A) \sqcup \Omega(-A)} |v_k(\mathbf{x}) - u(\mathbf{x}) + c|$$

$$= \min_{c \in \mathbb{R}} \max_{\mathbf{x} \in \Omega(A) \sqcup \Omega(-A)} \left| \sum_{j=0}^{k} (\theta_j B) \circ \sigma^j(\mathbf{x}) + c \right|.$$

On the one hand, for any $\mathbf{x} \in \Omega(A)$, we clearly have

$$\left| \sum_{j=0}^{k} (\theta_j B) \circ \sigma^j(\mathbf{x}) + c \right| = |c|.$$

On the other hand, since $\max_{\mathbf{x} \in \Omega(-A)} B(\mathbf{x}) =: -\gamma < 0$, for all $\mathbf{x} \in \Omega(-A)$, we obtain

$$\sum_{j=0}^{k} (\theta_j B) \circ \sigma^j(\mathbf{x}) \leq \eta S_{k+1} B(\mathbf{x}) \leq -\eta(k+1)\gamma,$$

which yields

$$\left| \sum_{j=0}^{k} (\theta_j B) \circ \sigma^j(\mathbf{x}) + c \right| \geq (k+1)\gamma\eta - |c|.$$

Therefore, we conclude that

$$\|[v_k] - [u]\|_\# \geq \min_{c \geq 0} \max\{c, \ (k+1)\gamma\eta - c\} = \frac{(k+1)\gamma\eta}{2},$$

which shows that $\|[v_k] - [u]\|_\# \to \infty$. $\square$

Since we have actually ensured the existence of Lipschitz continuous separating sub-actions, we may use the same arguments of the previous proof to obtain an analogous result for the Lipschitz category. More precisely, if A is not cohomologous to a constant, then, in the quotient space of Lipschitz continuous real-valued functions on Σ identified when their difference is identically constant, with respect to the topology of the norm $\|\cdot\|_\#$, Lipschitz continuous sub-actions for A also form a non-compact subset.

Corollary 8.1 *If the Lipschitz continuous potential A is not cohomologous to a constant, then the Lipschitz continuous separating sub-actions for A that cannot be represented as in (7.1) form a non-compact subset of the quotient space.*

Proof From Propositions 5.2 and 3.3, any Lipschitz continuous separating sub-action u_ρ given by (7.1) verifies $\mathrm{Lip}(u_\rho) \leq \frac{\lambda}{1-\lambda}\mathrm{Lip}(S_{K_0}A)$. Therefore, whereas Lipschitz continuous sub-actions form a non-compact subset, all u_ρ's together give rise to a compact subset of the quotient space. However, as we have pointed out, Lipschitz continuous separating sub-actions are dense among Lipschitz continuous sub-actions. $\square$

Despite its non-compactness, it is easy to exhibit a first extremal point of the convex set of continuous sub-actions. Since such a discussion in fact takes place in $(C(\Sigma)/\mathbb{R}, \|\cdot\|_\#)$, let us consider for a moment the set of continuous sub-actions as

$$\left\{u \in C(\Sigma) : u \text{ is a sub-action with } \max_{\Omega(A)} u = 0\right\}.$$

Using this identification, the calibrated sub-action

$$v_0 := \mathcal{F}_A(0) = \min_{\mathbf{x}\in\Omega(A)} \phi_A(\mathbf{x},\cdot) = \min_{\mathbf{x}\in\Omega(A)} h_A(\mathbf{x},\cdot)$$

is an extremal point. Indeed, note first that, for any point $\mathbf{y} \in \Omega(A)$, clearly

$$v_0(\mathbf{y}) = \min_{\mathbf{x}\in\Omega(A)} \phi_A(\mathbf{x},\mathbf{y}) \leq \phi_A(\mathbf{y},\mathbf{y}) = 0.$$

In particular, we get $\max_{\Omega(A)} v_0 \leq 0$. Let then $u \in C(\Sigma)$ be an arbitrary sub-action satisfying $\max_{\Omega(A)} u = 0$. For $\mathbf{x} \in \Omega(A)$ and $\mathbf{y} \in \Sigma$, thanks to item i of Proposition 5.2, we verify

$$u(\mathbf{y}) \leq u(\mathbf{x}) + \phi_A(\mathbf{x},\mathbf{y}) \leq \phi_A(\mathbf{x},\mathbf{y}).$$

Hence, it follows $u \leq v_0$, which ensures simultaneously that $\max_{\Omega(A)} v_0 = 0$ and v_0 is an extremal sub-action.

Another example of an extremal point of the convex set of continuous sub-actions can be pointed out if we now take into account the identification

$$\{u \in C(\Sigma) : u \text{ is a sub-action with } \max u = 0\}.$$

For $\mathbf{x} \in \Sigma$, denote

$$w_0(\mathbf{x}) := \inf_{k\geq 0} T^k_{A-\beta_A}(0)(\mathbf{x}) = \inf_{k\geq 0} \min_{\sigma^k(\bar{\mathbf{x}})=\mathbf{x}} [-S_k(A-\beta_A)(\bar{\mathbf{x}})].$$

Since by convention $S_0(A - \beta_A) = 0$, clearly $w_0(\mathbf{x}) \leq 0$. Note then that $u \in C(\Sigma)$ is a calibrated sub-action for A if, and only if, $T_{A-\beta_A}^k(u) = u$ for all $k \geq 1$. Thus, given a calibrated sub-action u, since T_A is a nonexpansive mapping, we have $T_{A-\beta_A}^k(0) = [T_A^k(0) - T_A^k(u)] + u \geq -2\|u\|_\infty$, which implies that $w_0(\mathbf{x}) \geq -2\|u\|_\infty$. We have shown that w_0 is a well-defined real-valued function on Σ, which a priori is upper semi-continuous and bounded. Furthermore, using that Lax-Oleinik operators commute with infima, we obtain

$$T_{A-\beta_A}(w_0) = \inf_{k \geq 1} T_{A-\beta_A}^k(0) \geq w_0,$$

which means that w_0 is a sub-action for A, but it may not be calibrated. Recalling the auxiliary function given by (5.1), notice that

$$w_0 = \inf_{\mathbf{x} \in \Sigma} \lim_{\epsilon \to 0} \mathfrak{S}_0^{\epsilon,0}(\mathbf{x}, \cdot) = \inf_{\mathbf{x} \in \Sigma} \lim_{\epsilon \to 0} \mathfrak{S}_0^{\epsilon,\epsilon}(\mathbf{x}, \cdot) \leq \inf_{\mathbf{x} \in \Omega(A)} \phi_A(\mathbf{x}, \cdot) = v_0.$$

Concerning its regularity, w_0 is actually a Lipschitz continuous function. The reader can see it by adapting arguments from Chaps. 3 or 5. A direct proof is as follows. Let $\mathbf{x}, \bar{\mathbf{x}} \in \Sigma$ be any points verifying $d(\mathbf{x}, \bar{\mathbf{x}}) = \lambda^n$ for some $n \geq 1$. Given $\epsilon > 0$, there are $\bar{\mathbf{y}} \in \Sigma$ and $\bar{n} \geq 0$, with $\sigma^{\bar{n}}(\bar{\mathbf{y}}) = \bar{\mathbf{x}}$, such that $w_0(\bar{\mathbf{x}}) + \epsilon > -S_{\bar{n}}A(\bar{\mathbf{y}})$. Consider $\mathbf{y} \in \Sigma$ with $d(\mathbf{y}, \bar{\mathbf{y}}) \leq \lambda^{n+\bar{n}}$ and $\sigma^{\bar{n}}(\mathbf{y}) = \mathbf{x}$, so that

$$w_0(\mathbf{x}) - w_0(\bar{\mathbf{x}}) - \epsilon < S_{\bar{n}}A(\bar{\mathbf{y}}) - S_{\bar{n}}A(\mathbf{y})$$

$$\leq \mathrm{Lip}(A)(\lambda^{n+\bar{n}} + \lambda^{n+\bar{n}-1} + \ldots + \lambda^n)$$

$$< \frac{\mathrm{Lip}(A)}{1-\lambda}\lambda^n = \frac{\mathrm{Lip}(A)}{1-\lambda}d(\mathbf{x}, \bar{\mathbf{x}}).$$

Since $\epsilon > 0$ can be taken arbitrarily small and $\mathbf{x}$ and $\bar{\mathbf{x}}$ play symmetric roles, we have

$$|w_0(\mathbf{x}) - w_0(\bar{\mathbf{x}})| \leq \frac{\mathrm{Lip}(A)}{1-\lambda}d(\mathbf{x}, \bar{\mathbf{x}}) \quad \text{whenever} \quad d(\mathbf{x}, \bar{\mathbf{x}}) \leq \lambda.$$

It remains to argue that w_0 is an extremal sub-action. Suppose that u is any continuous sub-action verifying $\max u = 0$. Given $\mathbf{x} \in \Sigma$, if the point $\mathbf{y} \in \Sigma$ satisfies $\sigma^n(\mathbf{y}) = \mathbf{x}$ for some $n \geq 0$, it is easy to see that $u(\mathbf{x}) \leq u(\mathbf{y}) - S_n(A - \beta_A)(\mathbf{y}) \leq -S_n(A - \beta_A)(\mathbf{y})$, which ensures that $u \leq w_0$. Besides, since $w_0 \leq 0$, evidently $\max u = 0$ implies $\max w_0 = 0$.

Chapter 9
Relations with the Thermodynamic Formalism

Perhaps one of the most interesting and fruitful applications of ergodic optimization theory occurs in the study of freezing phenomena in equilibrium statistical mechanics. In this concluding chapter, we provide a first glimpse of such a rich interaction among theories, by scrutinizing with basic techniques the convergence of equilibrium states to a particular maximizing probability on certain examples.

9.1 Equilibrium States and Maximizing Measures

Given a Lipschitz continuous potential $A : \Sigma \to \mathbb{R}$, via variational principle, the topological pressure is characterized by

$$P(A) = \max_{\mu \in \mathcal{M}_\sigma} \left[h_\mu(\sigma) + \int A \, d\mu \right],$$

where $h_\mu(\sigma)$ indicates the metric entropy. It is well known that this expression admits only one extremal probability $\mu_A \in \mathcal{M}_\sigma$, which is by definition the equilibrium state associated with the potential A.

The next proposition reveals interesting connections between equilibrium states and maximizing probabilities. In ergodic optimization, this result is part of the folklore, but a proof is reproduced for the reader's convenience. In equilibrium statistical mechanics, it corresponds to Aizemann-Lieb principle (see [4]) and the parameter t in its statement is interpreted as the inverse of the absolute temperature.

Proposition 9.1 *Let $A : \Sigma \to \mathbb{R}$ be a Lipschitz continuous potential. Then any weak* accumulation measure of the family $\{\mu_{tA}\}_{t>0}$ as t goes to infinity is an A-maximizing probability with metric entropy equal to*

$$\lim_{t \to \infty} h_{\mu_{tA}}(\sigma) = \lim_{t \to \infty} [P(tA) - t\beta_A] = \max_{\mu \in m_A} h_\mu(\sigma) = h_{top}(\sigma|_{\Omega(A)}),$$

© The Author(s) 2017
E. Garibaldi, *Ergodic Optimization in the Expanding Case*,
SpringerBriefs in Mathematics, DOI 10.1007/978-3-319-66643-3_9

which in particular means that such an accumulation probability realizes the topological entropy of σ restricted to the Aubry set.

Proof Let $t > 0$. On the one hand, for any measure $\nu \in \mathcal{M}_\sigma$,

$$t \int A \, d\nu \leq h_\nu(\sigma) + t \int A \, d\nu \leq h_{\mu_{tA}}(\sigma) + t \int A \, d\mu_{tA} \leq h_{\text{top}}(\sigma) + t \int A \, d\mu_{tA}.$$

On the other hand, if $\mu \in m_A$, we obtain

$$t \int A \, d\mu_{tA} \leq t \int A \, d\mu \leq h_{\text{top}}(\sigma) + t \int A \, d\mu.$$

In particular, for any A-maximizing probability μ, one has

$$\left| \int A \, d\mu_{tA} - \beta_A \right| = \left| \int A \, d\mu_{tA} - \int A \, d\mu \right| \leq \frac{1}{t} h_{\text{top}}(\sigma),$$

which shows the first claim. Note now that, for all A-maximizing probability μ and $t > 0$, obviously $h_\mu(\sigma) + t\beta_A \leq P(tA) \leq h_{\mu_{tA}}(\sigma) + t\beta_A$. Therefore,

$$\max_{\mu \in m_A} h_\mu(\sigma) \leq \liminf_{t \to \infty}[P(tA) - t\beta_A] \leq \limsup_{t \to \infty}[P(tA) - t\beta_A] \leq \liminf_{t \to \infty} h_{\mu_{tA}}(\sigma).$$

Nevertheless, since the map $\nu \in \mathcal{M}_\sigma \mapsto h_\nu(\sigma) \in [0, \infty)$ is upper semi-continuous, from the claim already proved, it follows that

$$\limsup_{t \to \infty} h_{\mu_{tA}}(\sigma) \leq \max_{\mu \in m_A} h_\mu(\sigma).$$

Besides, from Corollary 7.4, one gets $\max_{\mu \in m_A} h_\mu(\sigma) = h_{\text{top}}(\sigma|_{\Omega(A)})$. □

The question whether the family of equilibrium states $\{\mu_{tA}\}_{t>0}$ does converge as $t \to \infty$ has become an interesting research theme into the intersection of ergodic optimization and equilibrium statistical mechanics. For locally constant potentials, the weak* convergence of these probabilities was proved first in [25] by semi-algebraic techniques and then in [72] from a dynamical approach. Algorithmic viewpoints may be found in [31, 50]. The max-plus community [5, 6] has obtained similar results. Besides, with a more physics flavor, a generic result may be found in [88]. For the general Lipschitz case, positive results [10, 11] have been produced, as well as counterexamples [30, 42] have been discussed. A discontinuous example [100] for which the limit does not exist was already known. Recently in [16], for a particular class of potentials which are Lipschitz continuous or, more generally, of summable variation, the regions of convergence or divergence of equilibrium states were fully described in terms of the Peierls barrier. Finally, generalizations of this kind of study for countable alphabet subshifts of finite type have received increased interest (see, for instance [14, 15, 61, 67, 69, 84]).

9.2 Examples of Convergence of Equilibrium States

We would like to end these notes discussing a particular case at which the limit of equilibrium states when the system is frozen may be explicitly determined. We will consider here a potential that depends on finitely many coordinates and we will impose a special condition on the decomposition into transitive pieces of its Aubry set.

If $A : \Sigma \to \mathbb{R}$ is a locally constant potential, by passing to a higher block presentation of Σ, we assume that A depends on two coordinates. As we have seen in the proof of Theorem 4.3, for such a potential

$$\Omega(A) = \{\mathbf{x} \in \Sigma : \mathbf{N}(x_j, x_{j+1}) = 1 \text{ for all } j \geq 0\},$$

where $\mathbf{N}$ is the transition matrix given by $\mathbf{N}(i, j) = 1$ if, and only if, there exists an A-maximizing periodic probability supported on the orbit of a point of Σ of the form $(i, j, x_2, \ldots, x_{n-1}, i, j, x_2, \ldots, x_{n-1}, \ldots)$. Moreover, if $u \in C(\Sigma)$ is a separating sub-action for A, then it is easy to see that the function

$$v(x_0) := \max_{\substack{(x_1, x_2, \ldots) \in \Sigma \\ \mathbf{M}(x_0, x_1) = 1}} u(x_0, x_1, x_2, \ldots)$$

defines a sub-action for A and also verifies a separating property in the sense that $A(x_0, x_1) + v(x_1) - v(x_0) = \beta_A \Leftrightarrow \mathbf{N}(x_0, x_1) = 1$. Therefore, replacing $A(x_0, x_1)$ by $A(x_0, x_1) + v(x_1) - v(x_0) - \beta_A$, we suppose from now on that $\beta_A = 0$, $A(x_0, x_1) \leq 0$, and $A(x_0, x_1) = 0 \Leftrightarrow \mathbf{N}(x_0, x_1) = 1$. Notice that these assumptions can be made without loss of generality.

Lemma 9.2 *The Aubry set of a locally constant potential A admits the following disjoint decomposition into transitive pieces:*

$$\Omega(A) = \Sigma_1 \sqcup \Sigma_2 \sqcup \ldots \sqcup \Sigma_s,$$

where the restriction of σ to each subshift of finite type Σ_l is transitive.

Given $j_0, \ldots, j_k \in \{1, \ldots, r\}$, we will denote the associated cylinder set of a subshift of finite type $X \subset \{1, \ldots, r\}^{\mathbb{N}}$ by

$$[j_0, \ldots, j_k]_X := \{\mathbf{x} \in X : x_0 = j_0, \ldots, x_k = j_k\}.$$

Proof Fix any symbol $\iota_1 \in \{1, \ldots, r\}$ such that the cylinder set $[\iota_1]_{\Omega(A)}$ is nonempty. Define the transition matrix $\mathbf{N}_1$ by imposing that $\mathbf{N}_1(i, j) = 1$ if, and only if, $[\iota_1]_{\Omega(A)} \cap \sigma^{-k}([i, j]_{\Omega(A)}) \neq \emptyset$ for some $k \geq 0$. Notice that we have implicitly determined a sub-alphabet $\mathcal{A}_1 \subset \{1, \ldots, r\}$ which consists of all symbols appearing in an $\mathbf{N}$-allowed word $(w_0, w_1, \ldots, w_n)$ with $w_0 = w_n = \iota_1$. Consider thus

$$\Sigma_1 := \{\mathbf{x} \in \Omega(A) : \mathbf{N}_1(x_j, x_{j+1}) = 1 \text{ for all } j \geq 0\}.$$

By construction, the restriction of σ to Σ_1 is transitive. If there exists a symbol $\iota_2 \in \{1, \ldots, r\} \backslash \mathcal{A}_1$ such that the cylinder set $[\iota_2]_{\Omega(A)}$ is nonempty, we can inductively continue the decomposition of $\Omega(A)$ into transitive pieces. It is straightforward that such a decomposition will be disjoint: actually the sub-alphabets $\mathcal{A}_l$, $1 \leq l \leq s$, formed by the symbols that occur in sequences of Σ_l are two-by-two disjoint. $\square$

In the previous construction, the $r \times r$ transition matrix $\mathbf{N}_l$ that defines Σ_l is naturally identified with a $\#\mathcal{A}_l \times \#\mathcal{A}_l$ irreducible transition matrix. Committing an abuse of notation, we will also denote $\mathbf{N}_l$ this last one. Without loss of generality, by a simple renomination of the symbols of the original alphabet, we may always assume that $\mathbf{N}$ is a block diagonal matrix

$$\mathbf{N} = \mathrm{diag}(\mathbf{N}_1, \mathbf{N}_2, \ldots, \mathbf{N}_s, 0),$$

with $h_{\mathrm{top}}(\sigma|_{\Omega(A)}) = h_{\mathrm{top}}(\sigma|_{\Sigma_1}) \geq h_{\mathrm{top}}(\sigma|_{\Sigma_2}) \geq \ldots \geq h_{\mathrm{top}}(\sigma|_{\Sigma_s})$, where 0 indicates the $\left(r - \sum_{l=1}^{s} \#\mathcal{A}_l\right) \times \left(r - \sum_{l=1}^{s} \#\mathcal{A}_l\right)$ null matrix.

Before restricting our attention to a particular decomposition of the Aubry set, let us recall some well-known facts about equilibrium states associated with potentials that depend on two coordinates. For any $t > 0$, the $r \times r$ irreducible nonnegative matrix $\mathcal{L}_t$ defined by

$$\mathcal{L}_t(i,j) = \mathbf{M}(i,j)e^{tA(i,j)}, \qquad \forall\, i,j = 1, \ldots, r,$$

has as largest eigenvalue $e^{P(tA)}$. Denote respectively e^{V_t} and e^{W_t} the right and left strictly positive eigenvectors,

$$\sum_{j=1}^{r} \mathcal{L}_t(i,j)e^{V_t(j)} = e^{P(tA)}\, e^{V_t(i)} \quad \text{and} \quad \sum_{i=1}^{r} e^{W_t(i)}\mathcal{L}_t(i,j) = e^{P(tA)}\, e^{W_t(j)},$$

normalized by $\sum_{i=1}^{r} e^{V_t(i)} = \sum_{i=1}^{r} e^{W_t(i)} = 1$. Given then the stochastic matrix

$$\mathcal{Q}_t(i,j) = \mathcal{L}_t(i,j)e^{V_t(j)-V_t(i)-P(tA)}, \qquad \forall\, i,j = 1, \ldots, r,$$

let H_t be its strictly positive left eigenvector normalized by $\sum_{i=1}^{r} H_t(i) = 1$, more precisely: $H_t(i) = e^{V_t(i)+W_t(i)} / \sum_{k=1}^{r} e^{V_t(k)+W_t(k)}$, for all $i = 1, \ldots, r$. The equilibrium state μ_{tA} is thus the Markov probability on Σ defined by

$$\mu_{tA}([j_0, j_1, \ldots, j_n]_\Sigma) = H_t(j_0)\mathcal{Q}_t(j_0, j_1) \cdots \mathcal{Q}_t(j_{n-1}, j_n)$$

for any cylinder set $[j_0, j_1, \ldots, j_n]_\Sigma$. Therefore, the convergence (with respect to the Euclidean topology) of the family of pairs of vectors and matrices $(H_t, \mathcal{Q}_t)$ as t goes to infinity ensures the convergence of the family of probabilities μ_{tA} in this situation.

Proposition 9.3 *Suppose that the decomposition* $\Omega(A) = \Sigma_1 \sqcup \Sigma_2 \sqcup \ldots \sqcup \Sigma_s$ *is such that*

$$h_{top}(\sigma|_{\Omega(A)}) = h_{top}(\sigma|_{\Sigma_1}) = h_{top}(\sigma|_{\Sigma_2}) = \cdots = h_{top}(\sigma|_{\Sigma_s})$$

and $\mathcal{A}_1 \sqcup \mathcal{A}_2 \sqcup \ldots \sqcup \mathcal{A}_s = \{1, \ldots, r\}.$

Let e^{F_l} *denote a strictly positive right eigenvector of the* $\#\mathcal{A}_l \times \#\mathcal{A}_l$ *irreducible matrix* $\mathbf{N}_l$ *associated with its largest eigenvalue* $e^{h_{top}(\sigma|_{\Omega(A)})}$. *Let* $\mathcal{Q}_{\mathbf{N}_l}$ *be the stochastic matrix defined by* $\mathcal{Q}_{\mathbf{N}_l}(i,j) = \mathbf{N}_l(i,j)e^{F_l(j)-F_l(i)-h_{top}(\sigma|_{\Omega(A)})}$ *for all* $i, j \in \mathcal{A}_l$. *Then,*

$$\lim_{t\to\infty} \mathcal{Q}_t = \mathcal{Q}_{\mathbf{N}} := diag(\mathcal{Q}_{\mathbf{N}_1}, \mathcal{Q}_{\mathbf{N}_2}, \ldots, \mathcal{Q}_{\mathbf{N}_s}).$$

Proof Let e^{F_l} and e^{G_l} be respectively the right and left strictly positive eigenvectors of $\mathbf{N}_l$ for its largest eigenvalue $e^{h_{top}(\sigma|_{\Omega(A)})}$, namely,

$$\sum_{j \in \mathcal{A}_l} \mathbf{N}_l(i,j)e^{F_l(j)} = e^{h_{top}(\sigma|_{\Omega(A)})} e^{F_l(i)} \quad \forall\, i \in \mathcal{A}_l \qquad \text{and}$$

$$\sum_{i \in \mathcal{A}_l} e^{G_l(i)} \mathbf{N}_l(i,j) = e^{h_{top}(\sigma|_{\Omega(A)})} e^{G_l(j)} \quad \forall\, j \in \mathcal{A}_l,$$

normalized by $\sum_{i \in \mathcal{A}_l} e^{F_l(i)} = \sum_{i \in \mathcal{A}_l} e^{G_l(i)} = 1$. We will show that $\mathcal{Q}_t(i,j)$ converges to $\mathbf{N}_l(i,j)e^{F_l(j)-F_l(i)-h_{top}(\sigma|_{\Omega(A)})} = \mathcal{Q}_{\mathbf{N}_l}(i,j)$ whenever $i, j \in \mathcal{A}_l$ and to 0 otherwise. The main point in our reasoning is the fact that the normalization of the potential A means that $\mathcal{L}_t(i,j) = 1$ if, and only if, $\mathbf{N}(i,j) = 1$.

On the one hand, since e^{V_t} is a left eigenvector of $\mathcal{L}_t$ for $e^{P(tA)}$, we have

$$\sum_{\substack{1 \le j \le r \\ i \in \mathcal{A}_l}} e^{G_l(i)} \mathcal{L}_t(i,j)e^{V_t(j)} = e^{P(tA)} \sum_{i \in \mathcal{A}_l} e^{G_l(i)+V_t(i)}.$$

On the other hand, the fact that $A(i,j) = 0 \Leftrightarrow \mathbf{N}(i,j) = 1$ gives us

$$\sum_{\substack{1 \le j \le r \\ i \in \mathcal{A}_l}} e^{G_l(i)} \mathcal{L}_t(i,j)e^{V_t(j)} = \sum_{i,j \in \mathcal{A}_l} e^{G_l(i)} \mathbf{N}_l(i,j)e^{V_t(j)} + \sum_{\substack{i \in \mathcal{A}_l \\ \mathbf{N}_l(i,j)=0}} e^{G_l(i)} \mathcal{L}_t(i,j)e^{V_t(j)}.$$

Since e^{G_l} is a left eigenvector of $\mathbf{N}_l$ for $e^{h_{top}(\sigma|_{\Omega(A)})}$, we thus obtain

$$\sum_{\substack{i \in \mathcal{A}_l \\ \mathbf{N}_l(i,j)=0}} e^{G_l(i)} \mathcal{L}_t(i,j)e^{V_t(j)} = \left(e^{P(tA)} - e^{h_{top}(\sigma|_{\Omega(A)})}\right) \sum_{i \in \mathcal{A}_l} e^{G_l(i)+V_t(i)}, \tag{9.1}$$

which yields

$$\sum_{i\in\mathcal{A}_l}\frac{e^{G_l(i)+V_t(i)}}{\sum_{k\in\mathcal{A}_l}e^{G_l(k)+V_t(k)}}\sum_{\mathbf{N}_l(i,j)=0}\mathcal{Q}_t(i,j)=1-e^{h_{\mathrm{top}}(\sigma|_{\Omega(A)})-P(tA)}. \tag{9.2}$$

Claim For all $i,j\in\mathcal{A}_l$, we have $e^{-\#\mathcal{A}_l P(tA)}<e^{V_t(i)-V_t(j)}<e^{\#\mathcal{A}_l P(tA)}$.

Since $(\Sigma_{\mathbf{N}_l},\sigma)$ is a transitive subshift of finite type, there exists an $\mathbf{N}_l$-allowed word $(w_0,w_1,\ldots,w_n)$, with $n<\#\mathcal{A}_l$, such that $w_0=i$ and $w_n=j$. Therefore,

$$e^{V_t(j)}=\mathcal{L}_t(w_0,w_1)\mathcal{L}_t(w_1,w_2)\cdots\mathcal{L}_t(w_{n-1},w_n)e^{V_t(j)}$$

$$\leq\sum_{1\leq i_1,\ldots,i_n\leq r}\mathcal{L}_t(i,i_1)\mathcal{L}_t(i_1,i_2)\cdots\mathcal{L}_t(i_{n-1},i_n)e^{V_t(i_n)}=e^{nP(tA)}e^{V_t(i)},$$

from which $e^{-\#\mathcal{A}_l P(tA)}<e^{V_t(i)-V_t(j)}$. By symmetry, the claim is proved.

By Proposition 9.1, $\lim_{t\to\infty}P(tA)=h_{\mathrm{top}}(\sigma|_{\Omega(A)})$. Hence, fixing any $i_l\in\mathcal{A}_l$, the previous claim ensures that the term

$$\frac{e^{G_l(i)+V_t(i)}}{\sum_{k\in\mathcal{A}_l}e^{G_l(k)+V_t(k)}}=\frac{e^{G_l(i)}\,e^{V_t(i)-V_t(i_l)}}{\sum_{k\in\mathcal{A}_l}e^{G_l(k)}\,e^{V_t(k)-V_t(i_l)}},\qquad i\in\mathcal{A}_l,$$

is greater than some strictly positive constant for every $t>0$. Therefore, for $i\in\mathcal{A}_l$, we conclude from Eq. (9.2) that

$$\lim_{t\to\infty}\mathcal{Q}_t(i,j)=0\quad\text{whenever}\quad\mathbf{N}_l(i,j)=0.$$

It remains to argue that, if $\mathbf{N}_l(i,j)=1$, then $\mathcal{Q}_t(i,j)$ converges to $\mathcal{Q}_{\mathbf{N}_l}(i,j)$ too. However, in this case, such a convergence is equivalent to

$$\lim_{t\to\infty}e^{V_t(j)-V_t(i)}=e^{F_l(j)-F_l(i)},\qquad\forall i,j\in\mathcal{A}_l. \tag{9.3}$$

Fix $i,i_l\in\mathcal{A}_l$. For any $j\in\mathcal{A}_l$, from the above claim, we obtain that the family $\{V_t(j)-V_t(i_l)\}_{t>0}$ is bounded. Thus, let $\bar{V}(j)$ be any accumulation point of this family as $t\to\infty$. Since

$$e^{P(tA)}=\sum_{j=1}^{r}\mathcal{L}_t(i,j)e^{V_t(j)-V_t(i)}$$

$$=\sum_{j\in\mathcal{A}_l}\mathbf{N}_l(i,j)e^{V_t(j)-V_t(i)}+\sum_{\mathbf{N}_l(i,j)=0}\mathcal{L}_t(i,j)e^{V_t(j)-V_t(i)},$$

by considering a suitable subfamily, the passage to the limit gives

$$e^{h_{\mathrm{top}}(\sigma|_{\Omega(A)})}=\sum_{j\in\mathcal{A}_l}\mathbf{N}_l(i,j)e^{\bar{V}(j)-\bar{V}(i)}.$$

Nevertheless, this equality indicates that $e^{\bar{V}}$ is a right strictly positive eigenvector of $\mathbf{N}_l$ for the eigenvalue $e^{h_{\text{top}}(\sigma|_{\Omega(A)})}$. In particular, $e^{\bar{V}}$ is colinear to e^{F_l} and we have $e^{\bar{V}(j)-\bar{V}(i)} = e^{F_l(j)-F_l(i)}$. □

Because of Proposition 9.1, it would not be unreasonable to suspect the hypothesis $\mathcal{A}_1 \sqcup \ldots \sqcup \mathcal{A}_s = \{1, \ldots, r\}$ is unnecessary and to conjecture that a general conclusion should be $\lim_{t\to\infty} \mathcal{Q}_t = \text{diag}(\mathcal{Q}_{N_1}, \mathcal{Q}_{N_2}, \ldots, \mathcal{Q}_{N_s}, 0)$. Here is however a counterexample. Given the transition matrix $\mathbf{M} = \left(\begin{smallmatrix} 1 & 0 & 1 \\ 1 & 1 & 1 \\ 0 & 1 & 1 \end{smallmatrix} \right)$, suppose that $A(i,j) \leq 0$ whenever $\mathbf{M}(i,j) = 1$, and $A(i,j) = 0$ if, and only if, $i = j \in \{1,2\}$. Notice then that $\beta_A = 0$ and $\Omega(A) = \{\underline{1}, \underline{2}\}$, where $\underline{i}$ indicates the fixed point $(i,i,i,\ldots)$. Hence, in this case, $s = 2$ with $\mathcal{A}_1 = \{1\}$ and $\mathcal{A}_2 = \{2\}$. From the eigen-equation

$$
\begin{pmatrix} 1 & 0 & e^{tA(1,3)} \\ e^{tA(2,1)} & 1 & e^{tA(2,3)} \\ 0 & e^{tA(3,2)} & e^{tA(3,3)} \end{pmatrix} \begin{pmatrix} e^{V_t(1)-V_t(3)} \\ e^{V_t(2)-V_t(3)} \\ 1 \end{pmatrix} = e^{P(tA)} \begin{pmatrix} e^{V_t(1)-V_t(3)} \\ e^{V_t(2)-V_t(3)} \\ 1 \end{pmatrix},
$$

it is clear that $e^{V_t(2)-V_t(3)} = \left(e^{P(tA)} - e^{tA(3,3)} \right) / e^{tA(3,2)}$. In particular, one has $\mathcal{Q}_t(3,2) = 1 - e^{tA(3,3)}/e^{P(tA)} \to 1$ as $t \to \infty$, which contradicts such a conjecture.

We will now present a result on the convergence of the family of vectors H_t, which together with the previous proposition will ensure the convergence of the corresponding family of equilibrium states. In order to simplify the exposition, we will consider from now on the case $s = 2$. More precisely, we focus on a decomposition $\Omega(A) = \Sigma_1 \sqcup \Sigma_2$ such that $h_{\text{top}}(\sigma|_{\Omega(A)}) = h_{\text{top}}(\sigma|_{\Sigma_1}) = h_{\text{top}}(\sigma|_{\Sigma_2})$ and $\mathcal{A}_1 \sqcup \mathcal{A}_2 = \{1, \ldots, r\}$, with $1 \in \mathcal{A}_1$ and $r \in \mathcal{A}_2$. By conciseness, we write $\gamma_{ij} = \gamma_{ij}(t) := e^{V_t(i)-V_t(j)}$ and $\bar{\gamma}_{ij} = \bar{\gamma}_{ij}(t) := e^{W_t(i)-W_t(j)}$, so that

$$
H_t(i) = \frac{\gamma_{i1}\bar{\gamma}_{i1}}{\sum_{k\in\mathcal{A}_1} \gamma_{k1}\bar{\gamma}_{k1} + \left(\sum_{k\in\mathcal{A}_2} \gamma_{kr}\bar{\gamma}_{kr}\right)\gamma_{r1}\bar{\gamma}_{r1}} \quad \text{for } i \in \mathcal{A}_1 \quad \text{and}
$$

$$
H_t(i) = \frac{\gamma_{ir}\bar{\gamma}_{ir}}{\left(\sum_{k\in\mathcal{A}_1} \gamma_{k1}\bar{\gamma}_{k1}\right)\gamma_{1r}\bar{\gamma}_{1r} + \sum_{k\in\mathcal{A}_2} \gamma_{kr}\bar{\gamma}_{kr}} \quad \text{for } i \in \mathcal{A}_2.
$$

Notice that the limit (9.3) obtained during the proof of Proposition 9.3 may be recast as $\gamma_{ij} \to e^{F_l(i)-F_l(j)}$ for all $i,j \in \mathcal{A}_l$. Similarly, one can show that $\bar{\gamma}_{ij} \to e^{G_l(i)-G_l(j)}$ for all $i,j \in \mathcal{A}_l$. Therefore, from the above equations, it is easy to see that, in order to show the convergence of H_t in such a situation, it is enough to guarantee the convergence of $\gamma_{1r} = \gamma_{r1}^{-1}$ and of $\bar{\gamma}_{1r} = \bar{\gamma}_{r1}^{-1}$ as t goes to infinity. We will then prove a lemma that gives the asymptotic behavior of γ_{1r}. An analogous result may be obtained for $\bar{\gamma}_{1r}$.

First, notice that (9.1) may be rewritten according to $l = 1, 2$ as

$$
\sum_{i,j\in\mathcal{A}_1} [1 - \mathbf{N}_1(i,j)]e^{G_1(i)}\mathcal{L}_t(i,j)e^{V_t(i)} + \sum_{\substack{i\in\mathcal{A}_1 \\ j\in\mathcal{A}_2}} e^{G_1(i)}\mathcal{L}_t(i,j)e^{V_t(i)}
$$

$$
= \left(e^{P(tA)} - e^{h_{\text{top}}(\sigma|_{\Omega(A)})} \right) \sum_{i\in\mathcal{A}_1} e^{G_1(i)+V_t(i)},
$$

(9.4)

$$\sum_{\substack{i\in\mathcal{A}_2 \\ j\in\mathcal{A}_1}} e^{G_2(i)}\mathcal{L}_t(i,j)e^{V_t(j)} + \sum_{i,j\in\mathcal{A}_2}[1-\mathbf{N}_2(i,j)]e^{G_2(i)}\mathcal{L}_t(i,j)e^{V_t(j)}$$

$$= \left(e^{P(tA)} - e^{h_{\text{top}}(\sigma|_{\Omega(A)})}\right)\sum_{i\in\mathcal{A}_2} e^{G_2(i)+V_t(i)}. \tag{9.5}$$

Denote thus

$$\Lambda_{11} := \sum_{i,j\in\mathcal{A}_1}[1-\mathbf{N}_1(i,j)]\frac{e^{G_1(i)}\mathcal{L}_t(i,j)e^{V_t(j)}}{\sum_{k\in\mathcal{A}_1}e^{G_1(k)+V_t(k)}}, \tag{9.6}$$

$$\Lambda_{12} := \sum_{\substack{i\in\mathcal{A}_1 \\ j\in\mathcal{A}_2}} \frac{e^{G_1(i)}\mathcal{L}_t(i,j)e^{V_t(j)}}{\sum_{k\in\mathcal{A}_2}e^{G_2(k)+V_t(k)}}, \tag{9.7}$$

$$\Lambda_{21} := \sum_{\substack{i\in\mathcal{A}_2 \\ j\in\mathcal{A}_1}} \frac{e^{G_2(i)}\mathcal{L}_t(i,j)e^{V_t(j)}}{\sum_{k\in\mathcal{A}_1}e^{G_1(k)+V_t(k)}}, \tag{9.8}$$

$$\Lambda_{22} := \sum_{i,j\in\mathcal{A}_2}[1-\mathbf{N}_2(i,j)]\frac{e^{G_2(i)}\mathcal{L}_t(i,j)e^{V_t(j)}}{\sum_{k\in\mathcal{A}_2}e^{G_2(k)+V_t(k)}}, \tag{9.9}$$

$$\Gamma_1 := \sum_{k\in\mathcal{A}_1} e^{G_1(k)+V_t(k)} \quad \text{and} \quad \Gamma_2 := \sum_{k\in\mathcal{A}_2} e^{G_2(k)+V_t(k)}.$$

Hence, Eqs. (9.4) and (9.5) may be recast into a matrix form

$$\begin{pmatrix}\Lambda_{11} & \Lambda_{12} \\ \Lambda_{21} & \Lambda_{22}\end{pmatrix}\begin{pmatrix}\Gamma_1 \\ \Gamma_2\end{pmatrix} = \left(e^{P(tA)} - e^{h_{\text{top}}(\sigma|_{\Omega(A)})}\right)\begin{pmatrix}\Gamma_1 \\ \Gamma_2\end{pmatrix}. \tag{9.10}$$

In particular, $e^{P(tA)} - e^{h_{\text{top}}(\sigma|_{\Omega(A)})}$ is the positive root of the characteristic polynomial $x^2 - (\Lambda_{11} + \Lambda_{22})x + \Lambda_{11}\Lambda_{22} - \Lambda_{12}\Lambda_{21}$, namely,

$$e^{P(tA)} - e^{h_{\text{top}}(\sigma|_{\Omega(A)})} = \frac{\Lambda_{11} + \Lambda_{22} + \sqrt{(\Lambda_{11} - \Lambda_{22})^2 + 4\Lambda_{12}\Lambda_{21}}}{2}. \tag{9.11}$$

Besides, for

$$\Theta := \frac{\sum_{k\in\mathcal{A}_1} e^{G_1(k)+V_t(k)-V_t(1)}}{\sum_{k\in\mathcal{A}_2} e^{G_2(k)+V_t(k)-V_t(r)}}, \tag{9.12}$$

notice that $\Gamma_1/\Gamma_2 = \Theta\gamma_{1r}$, where $\Theta \to \frac{\sum_{k\in\mathcal{A}_1} e^{G_1(k)+F_1(k)-F_1(1)}}{\sum_{k\in\mathcal{A}_2} e^{G_2(k)+F_2(k)-F_2(r)}}$ as $t \to \infty$. Therefore, from (9.10), we obtain

$$\Lambda_{11} + \Lambda_{12}\Theta^{-1}\gamma_{r1} = e^{P(tA)} - e^{h_{\text{top}}(\sigma|_{\Omega(A)})} = \Lambda_{21}\Theta\gamma_{1r} + \Lambda_{22}.$$

Using these equations and (9.11), we may conclude the following.

Lemma 9.3 *Suppose that the decomposition* $\Omega(A) = \Sigma_{N_1} \sqcup \Sigma_{N_2}$ *is such that* $h_{top}(\sigma|_{\Omega(A)}) = h_{top}(\sigma|_{\Sigma_1}) = h_{top}(\sigma|_{\Sigma_2})$ *and* $\mathcal{A}_1 \sqcup \mathcal{A}_2 = \{1, \ldots, r\}$, *with* $1 \in \mathcal{A}_1$ *and* $r \in \mathcal{A}_2$. *Then*

$$e^{V_t(1) - V_t(r)} = \frac{\Lambda_{11} - \Lambda_{22} + \sqrt{(\Lambda_{11} - \Lambda_{22})^2 + 4\Lambda_{12}\Lambda_{21}}}{2\Theta\Lambda_{21}} \qquad and$$

$$e^{V_t(r) - V_t(1)} = \frac{\Lambda_{22} - \Lambda_{11} + \sqrt{(\Lambda_{22} - \Lambda_{11})^2 + 4\Lambda_{12}\Lambda_{21}}}{2\Theta^{-1}\Lambda_{12}},$$

where the Λ_{ij}*'s are given in* (9.6) *to* (9.9) *and* Θ *is defined by* (9.12).

From this lemma, one may deduce that there are constants $\tau > 0$ and $\upsilon \in \mathbb{R}$ such that $\gamma_{1r}(t)/\tau e^{t\upsilon} \to 1$ as $t \to \infty$. A similar reasoning shows that there also exist constants $\bar{\tau} > 0$ and $\bar{\upsilon} \in \mathbb{R}$ with $\bar{\gamma}_{1r}(t)/\bar{\tau}e^{t\bar{\upsilon}} \to 1$ as $t \to \infty$. Therefore, we have obtained the convergence of H_t in this case. The next theorem summarizes our discussion so far.

Theorem 9.5 *Given a potential A that depends on $m + 1$ coordinates, suppose that the decomposition of its Aubry set results in exactly two transitive pieces such that both are of maximal topological entropy and the associated sub-alphabets consist of all $\mathbf{M}$-allowed words of length m. Then, the family of equilibrium states μ_{tA} converges as $t \to \infty$ and the weak* limit may be explicitly determined. Furthermore, the limit*

$$\lim_{t \to \infty} \frac{1}{t} \log \left(e^{P(tA) - t\beta_A} - e^{h_{top}(\sigma|_{\Omega(A)})} \right) \qquad (9.13)$$

does exist.

Of course, the above convergence of equilibrium states is just a particular case of Brémont's theorem [25]. The exponential rate (9.13) is obviously a direct consequence of (9.11). This phenomenon has been reported before (see, for instance [10, 11, 87]). For a characterization of this kind of limit, see Theorem 3.2.3 in [57].

Concerning the convergence of equilibrium states, we will provide here a couple of examples of the applicability of the previous results. A richer description of limits of equilibrium states associated with locally constant potentials when the system is frozen may be found, for instance, in [50]. In the examples that follow, potentials $A(i, j)$ are supposed to be normalized, that is: $\beta_A = 0$, $A(i, j) \leq 0$ and $A(i, j) = 0 \Leftrightarrow$ $\mathbf{N}(i, j) = 1$. Moreover, by simplicity, we also assume that A defines a symmetric matrix. Since the convergence of the family of matrices Q_t is completely described by Proposition 9.3, in each case we will focus our attention only on the limit of H_t as $t \to \infty$. By the symmetric assumption, clearly $e^{V_t} = e^{W_t}$, so that we will need to know just the asymptotic behavior of γ_{ij}. Hence, for strictly positive scalar functions we write $f \sim g$ to indicate that $f(t)/g(t) \to 1$ as $t \to \infty$, and we extend this notation to vector-valued functions when each pair of coordinate functions satisfies the property.

Let us first analyze a situation in which the Aubry set has positive topological entropy. Consider the transition matrix $\mathbf{M} = \left(\begin{smallmatrix} 1 & 1 & 0 & 1 \\ 1 & 1 & 0 & 0 \\ 0 & 0 & 1 & 1 \\ 1 & 0 & 1 & 1 \end{smallmatrix} \right)$ and denote

$$
\mathcal{L}_t = \begin{pmatrix} 1 & 1 & 0 & e^{tA(1,4)} \\ 1 & e^{tA(2,2)} & 0 & 0 \\ 0 & 0 & 1 & 1 \\ e^{tA(1,4)} & 0 & 1 & e^{tA(4,4)} \end{pmatrix} =: \begin{pmatrix} 1 & 1 & 0 & a_t \\ 1 & b_t & 0 & 0 \\ 0 & 0 & 1 & 1 \\ a_t & 0 & 1 & c_t \end{pmatrix}.
$$

Without loss of generality, we suppose that $A(2,2) \geq A(4,4)$. Evidently, $\Omega(A) = \Sigma_{\mathbf{N}_1} \sqcup \Sigma_{\mathbf{N}_2}$, where $\mathbf{N}_1 = \left(\begin{smallmatrix} 1 & 1 & 0 & 0 \\ 1 & 0 & 0 & 0 \\ 0 & 0 & 0 & 0 \\ 0 & 0 & 0 & 0 \end{smallmatrix} \right)$ and $\mathbf{N}_2 = \left(\begin{smallmatrix} 0 & 0 & 0 & 0 \\ 0 & 0 & 0 & 0 \\ 0 & 0 & 1 & 1 \\ 0 & 0 & 1 & 0 \end{smallmatrix} \right)$. From lemma (9.3), a direct calculation gives us

$$
\gamma_{14} \sim \begin{cases} \left(b_t + \sqrt{b_t^2 + 4\varphi^2 a_t^2} \right)/2a_t & \text{if } A(2,2) > A(4,4) \\ \varphi & \text{if } A(2,2) = A(4,4) \end{cases},
$$

where $\varphi := (1 + \sqrt{5})/2$. Notice now that

$$
H_t \sim \frac{1}{(1 + \varphi^{-2})\gamma_{14}^2 + \varphi^2 + 1} (\gamma_{14}^2, \varphi^{-2}\gamma_{14}^2, \varphi^2, 1).
$$

Thus, as $t \to \infty$, up to a normalization factor, the vector H_t converges to

(i) $(\varphi^2, 1, 0, 0)$ when $A(2,2) > \max\{A(1,4), A(4,4)\}$;
(ii) $\left((2\varphi^2 + 1 + \sqrt{1 + 4\varphi^2})\varphi^2, 2\varphi^2 + 1 + \sqrt{1 + 4\varphi^2}, 2\varphi^4, 2\varphi^2 \right)$ when $A(1,4) = A(2,2) > A(4,4)$;
(iii) $(\varphi^2, 1, \varphi^2, 1)$ when $A(1,4) > \max\{A(2,2), A(4,4)\}$ or $A(2,2) = A(4,4)$.

These methods can also be adapted to beyond the setting of Theorem 9.5. For an example, consider the full shift on three symbols and denote

$$
\mathcal{L}_t = \begin{pmatrix} 1 & e^{tA(1,2)} & e^{tA(1,3)} \\ e^{tA(1,2)} & 1 & e^{tA(2,3)} \\ e^{tA(1,3)} & e^{tA(2,3)} & 1 \end{pmatrix} =: \begin{pmatrix} 1 & a_t & b_t \\ a_t & 1 & c_t \\ b_t & c_t & 1 \end{pmatrix},
$$

where, without loss of generality, we assume that $A(1,2) \geq A(1,3) \geq A(2,3)$. In this case, $\Omega(A) = \Sigma_{\mathbf{N}_1} \sqcup \Sigma_{\mathbf{N}_2} \sqcup \Sigma_{\mathbf{N}_3}$, with $\mathbf{N}_1 = \left(\begin{smallmatrix} 1 & 0 & 0 \\ 0 & 0 & 0 \\ 0 & 0 & 0 \end{smallmatrix} \right)$, $\mathbf{N}_2 = \left(\begin{smallmatrix} 0 & 0 & 0 \\ 0 & 1 & 0 \\ 0 & 0 & 0 \end{smallmatrix} \right)$ and $\mathbf{N}_3 = \left(\begin{smallmatrix} 0 & 0 & 0 \\ 0 & 0 & 0 \\ 0 & 0 & 1 \end{smallmatrix} \right)$. Despite the fact that Proposition 9.3 still applies, we are obviously outside the context of Lemma 9.3. Let us describe how the technique behind this lemma can be nevertheless useful for this situation too. Using the equation $e^{V_t(2)} = \frac{a_t}{e^{P(tA)} - 1} e^{V_t(1)} + \frac{c_t}{e^{P(tA)} - 1} e^{V_t(3)}$, notice that

$$
\begin{pmatrix} \frac{a_t^2}{e^{P(tA)}-1} & b_t + \frac{a_t c_t}{e^{P(tA)}-1} \\ b_t + \frac{a_t c_t}{e^{P(tA)}-1} & \frac{c_t^2}{e^{P(tA)}-1} \end{pmatrix} \begin{pmatrix} e^{V_t(1)} \\ e^{V_t(3)} \end{pmatrix} = \left(e^{P(tA)} - 1\right) \begin{pmatrix} e^{V_t(1)} \\ e^{V_t(3)} \end{pmatrix}. \tag{9.14}
$$

Therefore, if we want as before to determine the asymptotic behavior of γ_{13}, we need first to know the asymptotic behavior of $e^{P(tA)} - 1$. Considering then the irreducible transition matrix $\mathbf{M}' = \begin{pmatrix} 0 & 1 & 1 \\ 1 & 0 & 1 \\ 1 & 1 & 0 \end{pmatrix}$, we define the potential $A'(i,j) = A(i,j)$ whenever $\mathbf{M}'(i,j) = 1$. Thanks to the eigen-equation

$$
\begin{pmatrix} 0 & a_t & b_t \\ a_t & 0 & c_t \\ b_t & c_t & 0 \end{pmatrix} \begin{pmatrix} e^{V_t(1)} \\ e^{V_t(2)} \\ e^{V_t(3)} \end{pmatrix} = \left(e^{P(tA)} - 1\right) \begin{pmatrix} e^{V_t(1)} \\ e^{V_t(2)} \\ e^{V_t(3)} \end{pmatrix},
$$

we thus see that $e^{P(tA)} - 1 = e^{P(tA')}$. Since $\lim_{t\to\infty}[P(tA') - t\beta_{A'}] = h_{\text{top}}(\sigma|_{\Omega(A')})$, we conclude that $e^{P(tA)} - 1 \sim \exp\left(t\beta_{A'} + h_{\text{top}}(\sigma|_{\Omega(A')})\right)$. Lemma 9.3 can be now adapted to describe the asymptotic behavior of γ_{13} from (9.14). Let us hence present the results. Denoting again $\varphi = (1 + \sqrt{5})/2$, one may check that

$$
\gamma_{13} \sim \begin{cases} a_t/(b_t + c_t) & \text{if } A(1,2) > A(1,3) \\ a_t\left(1 + \sqrt{1 + 4\varphi^2}\right)/2(\varphi b_t + c_t) & \text{if } A(1,2) = A(1,3) > A(2,3) \\ 1 & \text{if } A(1,2) = A(1,3) = A(2,3) \end{cases},
$$

Since $1 = \gamma_{12}\left(\frac{a_t}{e^{P(tA)}-1} + \frac{c_t}{e^{P(tA)}-1}\gamma_{31}\right)$, discussing case by case, we can deduce that always $\gamma_{12} \sim 1$. Recall that here $H_t = \frac{1}{1+\gamma_{21}^2+\gamma_{31}^2}(1, \gamma_{21}^2, \gamma_{31}^2)$. Finally, up to a normalization factor, we see that as $t \to \infty$ the vector H_t converges to

(i) $(1, 1, 0)$ when $A(1,2) > A(1,3)$;

(ii) $\left(1, 1, 4\varphi^2/\left(1 + \sqrt{1 + 4\varphi^2}\right)^2\right)$ when $A(1,2) = A(1,3) > A(2,3)$;

(iii) $(1, 1, 1)$ when $A(1,2) = A(1,3) = A(2,3)$.

On the full shift on three symbols, the reader can find in [50] the complete convergence scenario of equilibrium states associated with a (not necessarily symmetric) potential $A(i,j)$ whose Aubry set consists of exactly three fixed points.

Appendix A
Bounded Measurable Sub-actions

In this appendix, the reader will find a proof that a generic continuous potential cannot have a bounded measurable sub-action. The point is that the existence of such a sub-action implies that maximizing measures are characterized as those whose support is contained in a particular closed set. But, according to [22], for a generic continuous potential, this kind of characterization does not hold.

Maximizing Sets

When examining maximizing probabilities in the context of general continuous potentials over general continuous transformations of a compact metric space, the notion of *maximizing set* may be useful. By a maximizing set for a potential, we mean a closed subset such that an invariant probability is a maximizing measure if, and only if, its support is contained in such a set. In Chap. 1, we have argued that the existence of a continuous sub-action provides an immediate example of a maximizing set: the level set $(A + u \circ \sigma - u)^{-1}(\beta_A)$. As noted in Chap. 4, by taking into account Atkinson's theorem [7], another example of a maximizing set due to existence of continuous sub-actions is the Aubry set. The next result states that, in a general setting, the existence of a bounded measurable sub-action is sufficient to ensure the existence of a maximizing set.

Proposition A.1 *Let $T : X \to X$ be a continuous transformation on a compact metric space X. Suppose that the continuous potential $A \in C(X)$ admits a bounded measurable sub-action u, that is, a bounded measurable function $u : X \to \mathbb{R}$ such that everywhere on X*

$$A + u \circ T - u \leq \beta_A.$$

Then there exists a maximizing set for A.

© The Author(s) 2017
E. Garibaldi, *Ergodic Optimization in the Expanding Case*,
SpringerBriefs in Mathematics, DOI 10.1007/978-3-319-66643-3

Proof By taking the infimum over all real-valued continuous functions that are greater or equal to $S_k(A - \beta_A) + u \circ T^k - u$, one obtains an upper semi-continuous function $f_k : X \to \mathbb{R}$ such that

$$S_k(A - \beta_A)(x) + u \circ T^k(x) - u(x) \leq f_k(x) \qquad \forall\, x \in X.$$

Obviously $f_k \leq 0$ and $f_k \leq S_k(A - \beta_A) + 2\|u\|_\infty$.

Consider then the upper semi-continuous function $g_k := \frac{1}{k} f_k$. Since $g_k \leq 0$, note that each set $g_k^{-1}(0)$ is a closed subset of X. For any T-invariant probability μ, note also that

$$\int A \, d\mu - \beta_A \leq \int g_k \, d\mu \leq \int A \, d\mu - \beta_A + \frac{2}{k} \|u\|_\infty. \tag{A.1}$$

Thus, since $\int g_k \, d\mu \leq 0$, if μ is an A-maximizing measure, the first inequality in (A.1) ensures that $\mathrm{supp}(\mu) \subset g_k^{-1}(0)$ for each k.

Conversely, suppose that μ is a T-invariant probability whose support lies on the closed set $\bigcap_k g_k^{-1}(0)$. From the second inequality in (A.1), by passing to the limit one has $\beta_A \leq \int A \, d\mu$, that is, μ is an A-maximizing measure.   □

Note that the same proof shows the existence of a maximizing set under weaker hypotheses. For instance, the conclusion also holds if we assume that there is a sequence of bounded measurable real-valued functions $\{u_k\}$, with $\|u_k\|_\infty / k \to 0$ as $k \to \infty$, such that $S_k A + u_k \circ T^k - u_k \leq \beta_A$ on X.

The existence of a maximizing set is, nevertheless, incompatible with certain situations. The fact that a potential has, let us say, a unique maximizing measure of full support obviously prevents that there is a maximizing set for this potential. This type of somehow pathological behavior was revealed to be typical in [22]. See theorem C there. See also Theorem 4.2 in [64]. We next present the precise statement.

Theorem (Bousch-Jenkinson) *Let $T : X \to X$ be a continuous transformation on a compact metric space X. Given any proper closed invariant subset Y and any invariant probability μ such that $\mathrm{supp}(\mu) \subset Y$, suppose that μ is a weak* limit of a sequence of periodic probabilities with disjoint supports from Y. Then, for a generic continuous potential $A \in C(X)$, every A-maximizing probability has full support.*

The above approximation assumption is actually the result from which Sigmund [97] derived that periodic probabilities are a weak* dense among invariant probabilities for dynamics satisfying the specification property. Therefore, for transitive expanding dynamical systems, an immediate consequence is thus the following corollary.

Corollary A.2 *Let $T : X \to X$ be a transitive expanding transformation on a compact metric space X. Then, a generic continuous potential does not even admit a bounded measurable sub-action.*

Bibliography

1. Addas-Zanata, S., Tal, F.A.: Maximizing measures for endomorphisms of the circle. Nonlinearity **21**, 2347–2359 (2008)
2. Addas-Zanata, S., Tal, F.A.: On maximizing measures of homeomorphisms on compact manifolds. Fundam. Math. **200**, 145–159 (2008)
3. Addas-Zanata, S., Tal, F.A.: Support of maximizing measures for typical C dynamics on compact manifolds. Discrete Contin. Dyn. Syst. Ser. A **26**, 795–804 (2010)
4. Aizenman, M., Lieb, E.H.: The third law of thermodynamics and the degeneracy of the ground state for lattice systems. J. Stat. Phys. **24**, 279–297 (1981)
5. Akian, M., Bapat, R., Gaubert, S.: Asymptotics of the Perron eigenvalue and eigenvector using max-algebra. C. R. Acad. Sci. Ser. I **327**, 927–932 (1998)
6. Akian, M., Bapat, R., Gaubert, S.: Min-plus methods in eigenvalue perturbation theory and generalised Lidskii-Visik-Ljusternik theorem (2006, preprint)
7. Atkinson, G.: Recurrence of co-cycles and random walks. J. Lond. Math. Soc. **13**, 486–488 (1976)
8. Baccelli, F., Cohen, G., Olsder, G.J., Quadrat, J.P.: Synchronization and Linearity: An Algebra for Discrete Event Systems. Wiley Series in Probability and Mathematical Statistics. Wiley, New York (1992)
9. Baraviera, A.T., Lopes, A.O., Thieullen, P.: A large deviation principle for equilibrium states of Hölder potentials: the zero temperature case. Stochastics Dyn. **6**, 77–96 (2006)
10. Baraviera, A.T., Leplaideur, R., Lopes, A.O.: Selection of measures for a potential with two maxima at the zero temperature limit. SIAM J. Appl. Dyn. Syst. **11**, 243–260 (2012)
11. Baraviera, A.T., Leplaideur, R., Lopes, A.O.: Ergodic optimization, zero temperature limits and the max-plus algebra. 29° Colóquio Brasileiro de Matemática. IMPA, Rio de Janeiro (2013)
12. Bernard, P.: Existence of $C^{1,1}$ critical sub-solutions of the Hamilton-Jacobi equation on compact manifolds. Ann. Sci. Ec. Norm. Supér. **40**, 445–452 (2007)
13. Biryuk, A., Gomes, D.A.: An introduction to the Aubry-Mather theory. São Paulo J. Math. Sci. **4**, 17–63 (2010)
14. Bissacot, R., Freire Júnior, R.S.: On the existence of maximizing probabilities for irreducible countable Markov shifts: a dynamical proof. Ergodic Theory Dyn. Syst. **34**, 1103–1115 (2014)
15. Bissacot, R., Garibaldi, E.: Weak KAM methods and ergodic optimal problems for countable Markov shifts. Bull. Braz. Math. Soc. New Ser. **41**, 321–338 (2010)

© The Author(s) 2017
E. Garibaldi, *Ergodic Optimization in the Expanding Case*,
SpringerBriefs in Mathematics, DOI 10.1007/978-3-319-66643-3

16. Bissacot, R., Garibaldi, E., Thieullen, P.: Zero-temperature phase diagram for double-well type potentials in the summable variation class. Ergodic Theory Dyn. Syst. (2016). doi:10.1017/etds.2016.57

17. Blokhuis, A., Wilbrink, H.A.: Alternative proof of Sine's theorem on the size of a regular polygon in $\mathbb{R}^k$ with the ℓ_∞-metric. Discrete Comput. Geom. **7**, 433–434 (1992)

18. Bochi, J., Zhang, Y.: Ergodic optimization of prevalent super-continuous functions. Int. Math. Res. Not. **2016**, 5988–6017 (2016)

19. Bousch, T.: Le poisson n'a pas d'arêtes. Ann. Inst. Henri Poincaré Probab. Stat. **36**, 489–508 (2000)

20. Bousch, T.: La condition de Walters. Ann. Sci. Ec. Norm. Supér. **34**, 287–311 (2001)

21. Bousch, T.: Un lemme de Mañé bilatéral. C. R. Math. **335**, 533–536 (2002)

22. Bousch, T., Jenkinson, O.: Cohomology classes of dynamically non-negative C^k functions. Invent. Math. **148**, 207–217 (2002)

23. Branco, F.M.: Sub-actions and maximizing measures for one-dimensional transformations with a critical point. Discrete Contin. Dyn. Syst. Ser. A **17**, 271–280 (2007)

24. Branton, S.D.: Sub-actions for Young towers. Discrete Contin. Dyn. Syst. Ser. A **22**, 541–556 (2008)

25. Brémont, J.: Gibbs measures at temperature zero. Nonlinearity **16**, 419–426(2003)

26. Brémont, J.: Finite flowers and maximizing measures for generic Lipschitz functions on the circle. Nonlinearity **19**, 813–828 (2006)

27. Bressaud, X., Quas, A.: Rate of approximation of minimizing measures. Nonlinearity **20**, 845–853 (2007)

28. Butkovič, P.: Max-algebra: the linear algebra of combinatorics? Linear Algebra Appl. **367**, 313–335 (2003)

29. Butkovič, P., Schneider, H., Sergeev, S.: On visualization scaling, subeigenvectors and Kleene stars in max algebra. Linear Algebra Appl. **431**, 2395–2406 (2009)

30. Chazottes, J.R., Hochman, M.: On the zero-temperature limit of Gibbs states. Commun. Math. Phys. **297**, 265–281 (2010)

31. Chazottes, J.R., Gambaudo, J.M., Ugalde, E.: Zero-temperature limit of one-dimensional Gibbs states via renormalization: the case of locally constant potentials. Ergodic Theory Dyn. Syst. **31**, 1109–1161 (2011)

32. Chou, W., Griffiths, R.B.: Effective potentials, a new approach and new results for one-dimensional systems with competing length scales. Phys. Rev. Lett. **56**, 1929–1931 (1986)

33. Chou, W., Griffiths, R.B.: Ground states of one-dimensional systems using effective potentials. Phys. Rev. B **34**, 6219–6234 (1986)

34. Contreras, G.: Action potential and weak KAM solutions. Calc. Var. Partial Differ. Equ. **13**, 427–458 (2001)

35. Contreras, G.: Ground states are generically a periodic orbit. Invent. Math. **205**, 383–412 (2016)

36. Contreras, G., Iturriaga, R.: Global minimizers of autonomous Lagrangians. 22° Colóquio Brasileiro de Matemática. IMPA, Rio de Janeiro (1999)

37. Contreras, G., Delgado, J., Iturriaga, R.: Lagrangian flows: the dynamics of globally minimizing orbits II. Boletim da Sociedade Brasileira de Matemática **28**, 155–196 (1997)

38. Contreras, G., Iturriaga, R., Paternain, G.P., Paternain, M.: Lagrangian graphs, minimizing measures and Mañé's critical values. Geom. Funct. Anal. **8**, 788–809 (1998)

39. Contreras, G., Lopes, A.O., Thieullen, P.: Maximizing measures for expanding transformations (1998, preprint)

40. Contreras, G., Lopes, A.O., Thieullen, P.: Lyapunov minimizing measures for expanding maps of the circle. Ergodic Theory Dyn. Syst. **21**, 1379–1409 (2001)

41. Conze, J.P., Guivarc'h, Y.: Croissance des sommes ergodiques et principe variationnel. circa (1993, preprint)

42. Coronel, D., Rivera-Letelier, J.: Sensitive dependence of Gibbs measures at low temperatures. J. Stat. Phys. **160**, 1658–1683 (2015)

43. Cuninghame-Green, R.A.: Minimax algebra and applications. In: Hawkes, P.W. (ed.) Advances in Imaging and Electron Physics, vol. 90, pp. 1–121. Academic, San Diego (1995)

44. Fathi, A.: Théorème KAM faible et théorie de Mather sur les systèmes lagrangiens. C. R. Acad. Sci. Sér. I Math. **324**, 1043–1046 (1997)

45. Fathi, A.: The weak KAM theorem in Lagrangian dynamics. Cambridge Studies in Advanced Mathematics, vol. 88. Cambridge University Press, Cambridge (2016)

46. Fathi, A., Siconolfi, A.: Existence of C^1 critical subsolutions of the Hamilton-Jacobi equation. Invent. Math. **155**, 363–388 (2004)

47. Floría, L.M., Griffiths, R.B.: Numerical procedure for solving a minimization eigenvalue problem. Numer. Math. **55**, 565–574 (1989)

48. Garibaldi, E., Gomes, J.T.A.: Otimização de médias sobre grafos orientados. 29° Colóquio Brasileiro de Matemática. IMPA, Rio de Janeiro (2013)

49. Garibaldi, E., Lopes, A.O.: On the Aubry-Mather theory for symbolic dynamics. Ergodic Theory Dyn. Syst. **28**, 791–815 (2008)

50. Garibaldi, E., Thieullen, P.: Description of some ground states by Puiseux techniques. J. Stat. Phys. **146**, 125–180 (2012)

51. Garibaldi, E., Lopes, A.O., Thieullen, P.: On calibrated and separating sub-actions. Bull. Braz. Math. Soc. **40**, 577–602 (2009)

52. Garibaldi, E., Petite, S., Thieullen, P.: Calibrated configurations for Frenkel-Kontorova type models in almost-periodic environments. Ann. Henri Poincaré. **18**, 2905–2943 (2017)

53. Gaubert, S.: Théorie des systèmes linéaires dans les dioïdes. PhD thesis, École des Mines de Paris (1992)

54. Gaubert, S.: Methods and applications of (max, +) linear algebra. In: Reischuk, R., Morvan, M. (eds.) STACS 97, Proceedings of the 14th Annual Symposium on Theoretical Aspects of Computer Science Held in Lübeck 1997. Lecture Notes in Computer Science, vol. 1200, pp. 261–282. Springer, Berlin (1997)

55. Goebel, K., Kirk, W.A.: Iteration processes for nonexpansive mappings. In: Singh, S.P., Thomeier, S., Watson, B. (eds.) Topological Methods in Nonlinear Functional Analysis. Proceedings of Special Session on Fixed Point Theory and Applications Held in Toronto 1982, Contemporary Mathematics, vol. 21, pp. 115–123. AMS, Providence (1983)

56. Gomes, D.A.: Viscosity solutions of Hamilton-Jacobi equations. 27° Colóquio Brasileiro de Matemática. IMPA, Rio de Janeiro (2009)

57. Gomes, J.T.A.: Formalismos gibbsianos para sistemas de spins unidimensionais. Master's thesis, University of Campinas (2012)

58. Gondran, M., Minoux, M.: Graphs, dioïds and semirings: new models and algorithms. Operations Research/Computer Science Interfaces Series, vol. 41. Springer, Berlin (2008)

59. Heidergott, B., Olsder, G.J., van der Woude, J.W.: Max plus at work: modeling and analysis of synchronized systems: a course on max-plus algebra and its applications. Princeton Series in Applied Mathematics. Princeton University Press, Princeton (2006)

60. Hunt, B.R., Yuan, G.C.: Optimal orbits of hyperbolic systems. Nonlinearity **12**, 1207–1224 (1999)

61. Iommi, G., Yayama, Y.: Zero temperature limits of Gibbs states for almost-additive potentials. J. Stat. Phys. **155**, 23–46 (2014)

62. Ishikawa, S.: Fixed points and iteration of a nonexpansive mapping in a Banach space. Proc. Am. Math. Soc. **59**, 65–71 (1976)

63. Jenkinson, O.: Rotation, entropy, and equilibrium states. Trans. Am. Math. Soc. **353**, 3713–3739 (2001)

64. Jenkinson, O.: Ergodic optimization. Discrete Contin. Dyn. Syst. Ser. A **15**, 197–224 (2006)

65. Jenkinson, O.: Every ergodic measure is uniquely maximizing. Discrete Contin. Dyn. Syst. Ser. A **16**, 383–392 (2006)

66. Jenkinson, O.: A partial order on ×2-invariant measures. Math. Res. Lett. **15**, 893–900 (2008)

67. Jenkinson, O., Mauldin, R.D. Urbański, M.: Zero temperature limits of Gibbs-equilibrium states for countable alphabet subshifts of finite type. J. Stat. Phys. **119**, 765–776 (2005)

68. Karp, R.M.: A characterization of the minimum cycle mean in a digraph. Discrete Math. **23**, 309–311 (1978)
69. Kempton, T.: Zero temperature limits of Gibbs equilibrium states for countable Markov shifts. J. Stat. Phys. **143**, 795–806 (2011)
70. Kohlenbach, U.: A quantitative version of a theorem due to Borwein-Reich-Shafrir. Numer. Funct. Anal. Optim. **22**, 641–656 (2001)
71. Lemmens, B., Sheutzow, M.: On the dynamics of sup-norm non-expansive maps. Ergodic Theory Dyn. Syst. **25**, 861–871 (2005)
72. Leplaideur, R.: A dynamical proof for the convergence of Gibbs measures at temperature zero. Nonlinearity **18**, 2847–2880 (2005)
73. Livšic, A.N.: Homology properties of Y-systems. Mat. Zametki **10**, 758–763 (1971)
74. Lo, S.K.: Abschätzungen für starre mengen in $\mathbb{R}^l$ mit ℓ_∞-oder polygonaler norm. Master's thesis, University of Göttingen (1989)
75. Lopes, A.O., Thieullen, P.: Sub-actions for Anosov diffeomorphisms. Astérisque **287**, 135–146 (2003)
76. Lopes, A.O., Thieullen, P.: Sub-actions for Anosov flows. Ergodic Theory Dyn. Syst. **25**, 605–628 (2005)
77. Lopes, A.O., Rosas, V., Ruggiero, R.O.: Cohomology and subcohomology for expansive geodesic flows. Discrete Contin. Dyn. Syst. Ser. A **17**, 403–422 (2007)
78. Lopes Filho, M.C., Nussenzveig Lopes, H.J.: Uma introdução a soluções de viscosidade para equações de Hamilton-Jacobi. Monografias de Matemática. IMPA, Rio de Janeiro (1997)
79. Lyons, R.N., Nussbaum, R.D.: On transitive and commutative finite groups of isometries. In: Tan, K.K. (ed.) Fixed Point Theory and Applications, Proceedings of the Second International Conference on Fixed Point Theory and Applications Held in Halifax 1991, pp. 189–228. World Scientific, Singapore (1992)
80. Mañé, R.: Lagrangian flows: the dynamics of globally minimizing orbits. Bol. Soc. Brasil. Mat. **28**, 141–153 (1997)
81. Mather, J.N.: Variational construction of connecting orbits. Ann. Inst. Fourier **43**, 1349–1386 (1993)
82. Martus, P.: Asymptotic properties of nonstationary operator sequences in the nonlinear case. PhD thesis, Friedrich-Alexander University Erlangen-Nürnberg (1989)
83. Morris, I.D.: A sufficient condition for the subordination principle in ergodic optimization. Bull. Lond. Math. Soc. **39**, 214–220 (2007)
84. Morris, I.D.: Entropy for zero-temperature limits of Gibbs-equilibrium states for countable-alphabet subshifts of finite type. J. Stat. Phys. **126**, 315–324 (2007)
85. Morris, I.D.: Maximizing measures of generic Hölder functions have zero entropy. Nonlinearity **21**, 993–1000 (2008)
86. Morris, I.D.: The Mañé-Conze-Guivarc'h lemma for intermittent maps of the circle. Ergodic Theory Dyn. Syst. **29**, 1603–1611 (2009)
87. Morris, I.D.: A note on approximating the maximum ergodic average via the Ruelle pressure functional (2009, preprint)
88. Nekhoroshev, N.N.: Asymptotics of Gibbs measures in one-dimensional lattice models. Mosc. Univ. Math. Bull. **59**, 10–15 (2004)
89. Nussbaum, R.D.: Omega limit sets of nonexpansive maps: finiteness and cardinality estimates. Differ. Integr. Equ. **3**, 523–540 (1990)
90. Nussbaum, R.D.: Convergence of iterates of a nonlinear operator arising in statistical mechanics. Nonlinearity **4**, 1223–1240 (1991)
91. Parthasarathy, K.R.: On the category of ergodic measures. Ill. J. Math. **5**, 648–656 (1961)
92. Peres, Y.: A combinatorial application of the maximal ergodic theorem. Bull. Lond. Math. Soc. **20**, 248–252 (1988)
93. Pollicott, M., Sharp, R.: Livsic theorems, maximizing measures and the stable norm. Dyn. Syst. **19**, 75–88 (2004)
94. Quas, A., Siefken, J.: Ergodic optimization of super-continuous functions on shift spaces. Ergodic Theory Dyn. Syst. **32**, 2071–2082 (2012)

95. Radu, L.: Duality in thermodynamic formalism. circa (2004, preprint)
96. Savchenko, S.V.: Cohomological inequalities for finite Markov chains. Funct. Anal. Appl. **33**, 236–238 (1999)
97. Sigmund, K.: Generic properties of invariant measures for Axiom A diffeomorphisms. Invent. Math. **11**, 99–109 (1970)
98. Sine, R.: A nonlinear Perron-Frobenius theorem. Proc. Am. Math. Soc. **109**, 331–336 (1990)
99. Souza, R.R.: Sub-actions for weakly hyperbolic one-dimensional systems. Dyn. Syst. **18**, 165–179 (2003)
100. van Enter, A.C.D., Ruszel, W.M.: Chaotic temperature dependence at zero temperature. J. Stat. Phys. **127**, 567–573 (2007)
101. Walters, P.: An Introduction to Ergodic Theory. Graduate Texts in Mathematics, vol. 79. Springer, Berlin (1982)
102. Weller, D.: Hilbert's metric, part metric and selfmappings of a cone. PhD thesis, University of Bremen (1987)

Index

A
Aubry point, 21
Aubry set, 6, 21

C
calibrated sub-action, 5

E
equilibrium state, 53
ergodic maximizing value, 3
expanding dynamics, 3
extremal sub-action, 7, 50, 51

K
Kleene star, 35

L
Lax-Oleinik operator, 13
locally constant potential, 17

M
Mañé potential, 27
Mañé's critical value, 32

Mañé-Peierls transform, 37
max algebra, 34
max-plus algebra, 34
max-times algebra, 35
maximizing probability, 3
maximizing set, 65

O
optimal trajectory, 22

P
Peierls barrier, 6, 27
periodic probability, 9
potential, 3

S
separating sub-action, 6, 41
sub-action, 4
subshift of finite type, 3

T
topologically mixing shift, 3

© The Author(s) 2017
E. Garibaldi, *Ergodic Optimization in the Expanding Case*,
SpringerBriefs in Mathematics, DOI 10.1007/978-3-319-66643-3

Printed in the United States
By Bookmasters